MYWORKBOOK

GEX, INCORPORATED

PREALGEBRA
FOURTH EDITION

Tom Carson

PEARSON

Boston   Columbus   Indianapolis   New York   San Francisco   Upper Saddle River
Amsterdam   Cape Town   Dubai   London   Madrid   Milan   Munich   Paris   Montreal   Toronto
Delhi   Mexico City   Sao Paulo   Sydney   Hong Kong   Seoul   Singapore   Taipei   Tokyo

Reproduced by Pearson from electronic files supplied by the author.

ISBN-13: 978-0-321-78293-9
ISBN-10: 0-321-78293-3

1 2 3 4 5 6 BRR 15 14 13 12 11

www.pearsonhighered.com

PEARSON

Table of Contents

Chapter 1……………………………………………………………………………1

Chapter 2……………………………………………………………………..…29

Chapter 3……………………………………………………………………..…51

Chapter 4……………………………………………………………………..79

Chapter 5……………………………………………………………………100

Chapter 6……………………………………………………………………167

Chapter 7……………………………………………………………………203

Chapter 8……………………………………………………………………228

Chapter 9……………………………………………………………………246

Odd Answers………………………………………………………………..272

Chapter 1 WHOLE NUMBERS

1.1 Introduction to Numbers, Notation, and Rounding

Learning Objectives
1   Name the digit in a specified place.
2   Write whole numbers in standard and expanded form.
3   Write the word name for a whole number.
4   Graph a whole number on a number line.
5   Use <, >, or = to write a true statement.
6   Round numbers.
7   Interpret bar graphs and line graphs.

Key Terms

Use the vocabulary terms listed below to complete each statement in Exercises 1–7.

set	**subset**	**natural numbers**	**whole numbers**
equation	**inequality**	**statistic**	

1. The number 0 is an element of the _____.

2. A(n) _____ is a collection or group of elements.

3. A number used to describe a collection of numbers is a(n) _____.

4. The _____ are the counting numbers.

5. The mathematical relationship 2 < 12 is called a(n) _____.

6. The natural numbers are a(n) _____ of the whole numbers.

7. A statement used to show that two amounts are equal is a(n) _____.

GUIDED EXAMPLES AND PRACTICE

Objective 1 Name the digit in a specified place.

Review these examples for Objective 1:	Practice these exercises:
1. Identify the digit in the tens place; 5,833,801,265.	1. Identify the digit in the millions place; 7,838,902,215.
5,833,801,2**6**5 6	
2. Write the word name for the place value of the digit 4 in 51,468,098.	2. Write the word name for the place value of the digit 4 in 4,953,289,781.
51,**4**68,098 hundred thousands	

Objective 2 Write whole numbers in standard and expanded form.

Review these examples for Objective 2:

3. Write the number in expanded form; 61,932.

    6 ten thousands + 1 thousands + 9 hundreds + 3 tens + 2 ones

4. Write the number in standard form; 2 millions + 9 ten thousands + 7 thousands + 8 hundreds + 3 ones.

    2,097,803

Practice these exercises:

3. Write the number in expanded form; 35,609,972.

4. Write the number in standard form; 9 ten thousands + 2 hundreds + 3 tens + 8 ones.

Objective 3 Write the word name for a whole number.

Review this example for Objective 3:

5. Write the word name for 51,104.

    fifty-one thousand, one hundred four

Practice this exercise:

5. Write the word name for 416,995.

Objective 4 Graph a whole number on a number line.

Review this example for Objective 4:

6. Graph 3 on a number line.

    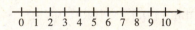

Practice this exercise:

6. Graph 5 on a number line.

Objective 5 Use <, >, or = to write a true statement.

Review this example for Objective 5:

7. Use <, >, or = to write a true statement.
    2672 _____ 1640

    2672   >   1640

Practice this exercise:

7. Use <, >, or = to write a true statement. 1891 _____ 2298

Objective 6 Round numbers.

Review this example for Objective 6:

8. Round 456,583,127 to the ten thousands place.

   A) Locate the digit in the ten thousands place, 8.
   B) The digit to the right is 3, which is less then 4; so round down.
   C) Change all digits to the right of 8 to zeros.

   456,580,000

Practice this exercise:

8. Round 456,583,127 to the hundreds place.

Objective 7 Interpret bar graphs and line graphs.

A garden club asked its members to vote for their favorite flower. The results are represented in the following graph. Use the graph to answer the following questions.

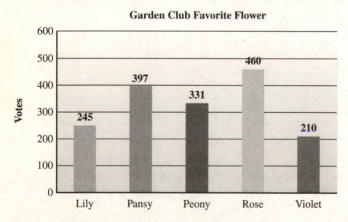

Garden Club Favorite Flower

Review this example for Objective 7:

9. Which flower received the least number of votes? Which flower received the most number of votes?

   The violet received the least and the rose received the most.

Practice this exercise:

9. Round the number of votes for peony to the nearest hundred.

ADDITIONAL EXERCISES
Objective 1 Name the digit in a specified place.
For extra help, see Example 1 on page 3 of your text and the Section 1.1 lecture video.
Identify the digit in the requested place.

1. 8,836,003,295; the hundred thousands place

2. 7,349,621; the thousands place

Write the word name for the place value of the digit 9 in each number.

3. 109

4. 37,492,561

Objective 2 Write whole numbers in standard and expanded form.
For extra help, see Examples 2–3 on pages 3–4 of your text and the Section 1.1 lecture video.
Write each number in expanded form.

5. 14,321

6. 720,516

Write each number in standard form.

7. 9 ten thousands + 2 hundreds + 3 tens + 8 ones

8. 5 hundred thousands + 3 thousands + 5 hundreds + 7 tens + 2 ones

Objective 3 Write the word name for a whole number.
For extra help, see Example 4 on page 4 of your text and the Section 1.1 lecture video.
Write the word name for each number.

9. 315,490

10. 2,568,157

Objective 4 Graph a whole number on a number line.

For extra help, see Example 5 on page 5 of your text and the Section 1.1 lecture video.

Graph each number on the number line.

11. 8

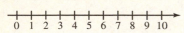

12. 1

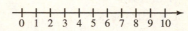

Objective 5 Use <, >, or = to write a true statement.

For extra help, see Example 6 on page 6 of your text and the Section 1.1 lecture video.

Use <, >, or = to write a true statement.

13. 562,104 _____ 562,104

14. 1,587,546 _____ 1,587,456

Objective 6 Round numbers.

For extra help, see Examples 7–8 on pages 7–8 of your text and the Section 1.1 lecture video.

Round 18,920,653 to the specified place.

15. millions

16. thousands

Objective 7 Interpret bar graphs and line graphs.

For extra help, see Examples 9–10 on pages 8–9 of your text and the Section 1.1 lecture video.

The yearly cost of tuition at a particular college, for five consecutive years, is represented in the following graph. Use the graph to answer the following questions.

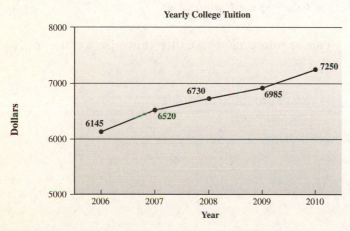

17. What was the cost of tuition in 2008? In 2006?

18. In what year was the cost of tuition the highest? What trend does the graph indicate about the cost of tuition?

Chapter 1 WHOLE NUMBERS

1.2 Adding and Subtracting Whole Numbers; Solving Equations

Learning Objectives
1 Add whole numbers.
2 Estimate sums.
3 Solve applications involving addition.
4 Subtract whole numbers.
5 Solve equations containing an unknown addend.
6 Solve applications involving subtraction.
7 Solve applications involving addition and subtraction.

Key Terms
Use the vocabulary terms listed below to complete each statement in Exercises 1–7.

addition   perimeter   variable   constant
subtraction   related equation   solution

1. The _____ is the value that makes an equation true when it replaces the variable in the equation.

2. To find the length of fence to put around a garden, find the _____ of the garden.

3. A(n) _____ is a symbol used to represent unknown amounts.

4. Given an addition equation involving an unknown amount, write a(n) _____ using subtraction to solve for the amount.

5. The operation _____ can be interpreted as a difference between amounts.

6. The operation _____ is use to combine amounts.

7. A symbol that does not change its value is a(n) _____.

GUIDED EXAMPLES AND PRACTICE

Objective 1 Add whole numbers.

Review this example for Objective 1:

1.   Add: 25,169 + 5,879 + 25

$$
\begin{array}{r}
{\scriptstyle 1\ \ 1\ \ 1\ \ 2}\\
2\,5,1\,6\,9\\
5,8\,7\,9\\
+2\,5\\
\hline
3\,1,0\,7\,3
\end{array}
$$

Practice this exercise:

1.
$$
\begin{array}{r}
64,891\\
+6,548\\
\hline
\end{array}
$$

Objective 2 Estimate sums.

Review this example for Objective 2:

2.   Estimate the sum by rounding. Then find the actual sum.

$$
\begin{array}{r}
7040\\
+2666\\
\hline
\end{array}
\qquad
\begin{array}{r}
7000\\
+3000\\
\hline
10,000
\end{array}
\leftarrow \text{Estimated sum.}
$$

$$
\begin{array}{r}
{\scriptstyle 1}\\
7040\\
+2666\\
\hline
9706
\end{array}
\leftarrow \text{Actual sum.}
$$

Practice this exercise:

2.   Estimate the sum by rounding. Then find the actual sum.

$$
\begin{array}{r}
76,037\\
7,017\\
+6,724\\
\hline
\end{array}
$$

Objective 3 Solve applications involving addition.

Review these examples for Objective 3:

3.   Solve. An actress waitresses part-time and spends the rest of her time trying to get an acting job. She has spent $1200 on acting classes, $55 on printing up resumes, and $150 on photos. How much has she spent trying to get an acting job?

Practice these exercises:

3.   Solve. Scott and Nancy want to add an addition to their house. The cost for materials will be $16,450, the labor will cost $8000, and the necessary permits and inspections will cost $2700. What will be the total price of the addition?

1

1200   Notice the phrase, *how much has*

 150   *she spent*, which tells us to add

+ 55   the amounts she has spent.
____

1405

She has spent $1305.

4. Find the perimeter of a 6-sided shape with side lengths of 12 feet, 10 feet, 14 feet, 8 feet, 2 feet and 16 feet.

The perimeter is the total distance around the shape. This is found by adding the lengths of the sides.

$P = 12 + 10 + 14 + 8 + 2 + 16$

$P = 62$

The perimeter is 62 ft.

4. Jim's yard is a 13-foot-long by 12-foot-long rectangle. What is the perimeter of his yard?

Objective 4 Subtract whole numbers.

Review this example for Objective 4:

5. Subtract.
$$\begin{array}{r} 859 \\ -178 \\ \hline \end{array}$$

$$\begin{array}{r} {}^{7}\;{}^{15}\\ \cancel{8}\cancel{5}9 \\ -178 \\ \hline 681 \end{array}$$

Practice this exercise:

5. Subtract.
$$\begin{array}{r} 758 \\ -303 \\ \hline \end{array}$$

Objective 5 Solve equations containing an unknown addend.

Review this example for Objective 5:

6. Solve and check. $8 + n = 14$

To solve for an unknown addend, write a related subtraction equation in which the known addend is subtracted from the sum.

$8 + n = 14$

$n = 14 - 8$ Check.

$n = 6$ $8 + 6 = 14$

Practice this exercise:

6. Solve and check.

$3 + r = 16$

8

Name: Date:
Instructor: Section:

Objective 6 Solve applications involving subtraction.

Review this example for Objective 6:

7.   Solve. Jan is making a costume for her son for the school pageant. She has 5 yards of material, and the costume requires 2 yards. How much of the material will Jan have left?

   The phrase, *how much of the material will Jan have left*, tells us to use subtraction. $5 - 2 = 3$

   Jan will have 3 yards left.

Practice this exercise:

7.   Solve. The attendance of a Thursday night baseball game is 52,366. The attendance on Friday is 53,933. How many more people attended Friday's game than Thursday's game?

Objective 7 Solve applications involving addition and subtraction.

Review this example for Objective 7:

8.   Solve. The diagram shows a circuit with two currents entering the node and two currents exiting the node. Find the missing current, in Amperes (A).

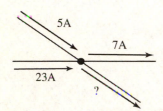

   First, calculate the total current entering the node.

   $23 + 5 = 28$

   Then, subtract the known current exiting the node from the total entering, to find the unknown current exiting.

   $28 - 7 = 21$

   The unknown current is 21A.

Practice this exercise:

8.   Solve. Josh has a storage capacity of 55 GB on his MP3 player. He has downloaded 9 GB of songs and 16 GB of movies. How much space does he have available?

ADDITIONAL EXERCISES
Objective 1 Add whole numbers.
For extra help, see Examples 1–2 on pages 13–14 of your text and the Section 1.2 lecture video.
Add.

1.
$$\begin{array}{r} 456 \\ +27 \\ \hline \end{array}$$

2.
$$\begin{array}{r} 161,256 \\ 278 \\ +4,218 \\ \hline \end{array}$$

3. $14 + 29 + 53$

4. $592 + 6375 + 1821$

Objective 2 Estimate sums.
For extra help, see Example 3 on page 15 of your text and the Section 1.2 lecture video.
Estimate each sum by rounding. Then find the actual sum.

5.
$$\begin{array}{r} 99,176 \\ 6,154 \\ +6,749 \\ \hline \end{array}$$

6. $8360 + 48 + 520 + 2854$

Objective 3 Solve applications involving addition.
For extra help, see Examples 4–5 on page 16 of your text and the Section 1.2 lecture video.
Solve.

7.   Oscar wants to cook dinner for a large family gathering. He estimates that he must spend $45 for the main course, $17 for side dishes, $31 for beverages, and $28 for dessert. What will be the total amount he spends on dinner?

8.   Find the perimeter of a 6-sided shape with side lengths of 12 feet, 10 feet, 14 feet, 8 feet, 2 feet and 16 feet.

Objective 4 Subtract whole numbers.
For extra help, see Example 6 on page 18 of your text and the Section 1.2 lecture video.
Subtract.

9.
$$\begin{array}{r} 5803 \\ -426 \\ \hline \end{array}$$

10. $85,801 - 12,424$

Objective 5 Solve equations containing an unknown addend.
For extra help, see Example 7 on page 19 of your text and the Section 1.2 lecture video.
Solve and check.

11. $66 + t = 122$

12. $146 + x = 284$

Objective 6 Solve applications involving subtraction.
For extra help, see Example 8 on page 20 of your text and the Section 1.2 lecture video.
Solve.

13. A roofer is replacing the roof on a house. He estimates it will require 240 squares of shingles to cover the whole roof. If he has 122 squares in stock, how many more squares will he need to order?

14. Emily has $200 in her savings account she has set aside for spending on new clothes. She heads to the mall and spends $55 on jeans, $38 on shoes, and $42 on a sweater. How much does she have left for accessories?

Objective 7 Solve applications involving addition and subtraction.
For extra help, see Example 9 on page 21 of your text and the Section 1.2 lecture video.
Solve.

15. A movie theater expects ticket sales of $9535 for the opening night of a new action movie. Typically, concession sales the opening night of a new film are about $1875. It will cost the theater $586 to pay employees that night, and concessions will cost the theater $265. How much will there be left from the ticket and concessions sales after these expenses are paid?

16. At midnight the outside temperature was 59 degrees Fahrenheit. The temperature rose 11 degrees over the next 8 hours, then dropped 3 degrees before noon. What was the temperature at noon?

Chapter 1 WHOLE NUMBERS

1.3 Multiplying Whole Numbers; Exponents

Learning Objectives
1   Multiply whole numbers.
2   Estimate products.
3   Solve applications involving multiplication.
4   Evaluate numbers in exponential form.
5   Write repeated factors in exponential form.
6   Solve applications.

Key Terms

Use the vocabulary terms listed below to complete each statement in Exercises 1–9.

> **multiplication   base   exponent   exponential form   rectangular array**
> **square unit   area   formula   distributive property**

1. To find how much carpet will be needed for a living room, find the
    _____ of the room.

2. The expression 15^3 is in _____.

3. The desks in a classroom are arranged in a(n) _____ if there is the
    same number of desks in each row.

4. The _____ of an exponential expression is the value that is used as a
    factor a number of times.

5. The operation _____ can be expressed as repeated addition.

6. The _____ may be used if a sum or difference is multiplied by a
    number.

7. A(n) _____ uses symbols in an equation to describe a procedure.

8. In an exponential expression, the _____ indicates the number of times
    the base is used as a factor.

9. A square that measures 1 unit by 1 unit is said to measure 1 _____.

GUIDED EXAMPLES AND PRACTICE

Objective 1   Multiply whole numbers.

Review this example for Objective 1:

1.   Multiply. $0(214)$

    The product of 0 and a number is 0.

    $0(214) = 0$

Practice this exercise:

1.   Multiply. $25 \cdot 1 \cdot 2$

Objective 2   Estimate products.

Review this example for Objective 2:

2.   Estimate the product by rounding. Then find the actual product. $13 \cdot 28$

    Round each factor to the highest possible place value so that each has only one nonzero digit. Then multiply the rounded factors.

    $10 \cdot 30 = 300$ ← Estimated product.

    $13 \cdot 28 = 364$ ← Actual product.

Practice this exercise:

2.   Estimate the product by rounding. Then find the actual product. 431×36

Objective 3   Solve applications involving multiplication.

Review this example for Objective 3:

3.   Solve. A restaurant has a dining room that is 99 feet by 42 feet. What is the area of the dining room?

    The key word, *by,* denotes multiplication.

    $99 \times 42 = 4158$

    The dining room is 4158 ft^2.

Practice this exercise:

3.   Solve. Lisa's car gets 20 miles per gallon of gasoline. How many miles can she drive on 53 gallons of gas?

Objective 4   Evaluate numbers in exponential form.

Review this example for Objective 4:

4.   Evaluate 2^7.

    To evaluate, write the base, 2, as a factor the number of times indicated by the exponent, 7, and then multiply.

    $2^7 = 2 \cdot 2 \cdot 2 \cdot 2 \cdot 2 \cdot 2 \cdot 2 = 128$

Practice this exercise:

4.   Evaluate 10^4.

Objective 5 Write repeated factors in exponential form.

Review these examples for Objective 5:

5.   Write $19 \cdot 19 \cdot 19 \cdot 19 \cdot 19$ in exponential form.

Since the number 19 is repeated, it is the base. The number of times it is repeated, 5, is the exponent.

$$19 \cdot 19 \cdot 19 \cdot 19 \cdot 19 = 19^5$$

6.   Write 17,544 in expanded form using powers of 10.

$$17,544 = 1 \times 10,000 + 7 \times 1,000 + 5 \times 100 + 4 \times 10 + 4 \times 1$$
$$= 1 \times 10^4 \quad + 7 \times 10^3 \quad + 5 \times 10^2 + 4 \times 10 + 4 \times 1$$

Practice these exercises:

5.   Write $3 \cdot 3 \cdot 3 \cdot 3 \cdot 3 \cdot 3 \cdot 3 \cdot 3 \cdot 3 \cdot 3 \cdot 3 \cdot 3$ in exponential form.

6.   Write 62,801,043 in expanded form using powers of 10.

Objective 6 Solve applications.

Review this example for Objective 6:

7.   Solve. A multiple-choice test consists of 5 questions with each question having 2 possible answers. How many different ways are there to answer the questions?

For each question there are 2 ways to answer and there are 5 questions. So the number of ways to answer the questions is $2 \cdot 2 \cdot 2 \cdot 2 \cdot 2 = 32$.

There are 32 different ways.

Practice this exercise:

7.   Solve. License plate tags in a particular state are to consist of 3 letters followed by 3 digits with repeated letters and digits allowed. How many different license plate tags can there be in this state?

ADDITIONAL EXERCISES
Objective 1 Multiply whole numbers.
For extra help, see Examples 1–2 on page 28 of your text and the Section 1.3 lecture video.
Multiply.

1. 103×608

2. 4116×692

Name: Date:
Instructor: Section:

Objective 2 Estimate products.

For extra help, see Example 3 on page 29 of your text and the Section 1.3 lecture video.
Estimate each product by rounding. Then find the actual product.

3. 439×816

4. 921×76

Objective 3 Solve applications involving multiplication.

For extra help, see Examples 4–6 on pages 29–31 of your text and the Section 1.3 lecture video.
Solve.

5. A bakery sells 120 cupcakes each day. How many cupcakes are sold in a week if the bakery is open every day?

6. If a carpenter uses 18 nails per cabinet, how many nails will he need to make 35 cabinets?

Objective 4 Evaluate numbers in exponential form.

For extra help, see Example 7 on page 33 of your text and the Section 1.3 lecture video.
Evaluate.

7. 1^9

8. 9^2

9. 7^3

10. 8^2

Objective 5 Write repeated factors in exponential form.

For extra help, see Examples 8–9 on page 34 of your text and the Section 1.3 lecture video.
Write in exponential form.

11. $21 \cdot 21 \cdot 21 \cdot 21$

12. $1 \cdot 1 \cdot 1 \cdot 1 \cdot 1 \cdot 1 \cdot 1$

13. $15 \cdot 15 \cdot 15$

14. $7 \cdot 7 \cdot 7 \cdot 7 \cdot 7 \cdot 7 \cdot 7 \cdot 7 \cdot 7$

Write in expanded form using powers of 10.

15. 785,463

16. 59,201

Objective 6 Solve applications.
For extra help, see Example 10 on page 36 of your text and the Section 1.3 lecture video.
Solve.

17. A rectangular room measures 21 feet wide by 18 feet long. How many 1 foot square tiles are needed to tile the floor?

18. A man is planning on staining his deck so, he needs to know what the area of the deck is. The deck measures 20 feet wide by 30 feet long. What is the area of the deck?

Chapter 1 WHOLE NUMBERS

1.4 Dividing Whole Numbers; Solving Equations

Learning Objectives
1   Divide whole numbers.
2   Solve equations containing an unknown factor.
3   Solve applications involving division.

Key Terms

Use the vocabulary terms listed below to complete each statement in Exercises 1–4.

> **division   remainder   divisor   dividend**

1. In the equation $24 \div 4 = 6$, 4 is the _____.

2. The operation of _____ can be thought of as repeated subtraction.

3. In the equation $15 \div 3 = 5$, 15 is the _____.

4. The amount left over after dividing two numbers is the _____.

GUIDED EXAMPLES AND PRACTICE

Objective 1 Divide whole numbers.

Review these examples for Objective 1:

1. Determine the quotient and explain your answer. $55 \div 55$

    $55 \div 55 = 1$ because $55 \times 1 = 55$

2. Use divisibility rules to determine whether 1230 is divisible by 2.

    If a number is even, then 2 is an exact divisor. A number is even if it has 0, 2, 4, 6, or 8 in the ones place. In the ones place of 1230 is 0, so it is even and therefore divisible by 2.

Practice this exercise:

1. Determine the quotient and explain your answer.
    $0 \div 8$

2. Use divisibility rules to determine whether 221 is divisible by 2.

3. Use divisibility rules to determine whether 727 is divisible by 3.

A number is divisible by 3 if the sum of the digits in the number is divisible by 3. The sum of the digits in 727 is $7 + 2 + 7 = 16$. Since 16 is not divisible by 3 then neither is 727.

4. Divide. $4452 \div 7$

$$
\begin{array}{r}
636 \\
7\overline{)4452} \\
-42 \\
\hline
25 \\
-21 \\
\hline
42 \\
-42 \\
\hline
0
\end{array}
$$

3. Use divisibility rules to determine whether 627 is divisible by 3.

4. Divide. $2960 \div 3$

Objective 2 Solve equations containing an unknown factor.

Review this example for Objective 2:

5. Solve and check. $2 \cdot y = 18$

To solve for an unknown factor, write a related division equation in which the product is divided by the known factor.

$\quad 2 \cdot y = 18$
$\qquad y = 18 \div 2 \qquad$ Check.
$\qquad y = 9 \qquad\qquad 2 \cdot 9 = 18$

Practice this exercise:

5. Solve and check. $8 \cdot x = 88$

Objective 3 Solve applications involving division.

Review this example for Objective 3:

6. Solve. A loan of $8784 will be paid off in 48 monthly payments. How much is each payment?

To solve, divide the loan amount into 48 equal payments.

Practice this exercise:

6. Solve. A bag can hold 22 kilograms of sand. How many bags can be filled with 1557 kilograms of sand? How many kilograms of sand will be left over?

$$\begin{array}{r} 183 \\ 48\overline{)8784} \\ -48 \\ \hline 398 \\ -384 \\ \hline 144 \\ -144 \\ \hline 0 \end{array}$$ Each payment is $183.

ADDITIONAL EXERCISES
Objective 1 Divide whole numbers.
For extra help, see Examples 1–6 on pages 41–44 of your text and the Section 1.4 lecture video.
Determine the quotient and explain your answer.

1. $27 \div 1$

2. $15 \div 0$

Use divisibility rules to determine whether the given number is divisible by 2.

3. 357

4. 170

Use divisibility rules to determine whether the given number is divisible by 3.

5. 953

6. 642

Use divisibility rules to determine whether the given number is divisible by 5.

7. 715

8. 5251

Divide.

9. $7158 \div 49$

10. $22,888 \div 57$

11. $\dfrac{80,000}{250}$

12. $\dfrac{4253}{34}$

Objective 2 Solve equations containing an unknown factor.
For extra help, see Example 7 on page 45 of your text and the Section 1.4 lecture video.
Solve and check.

13. $7 \cdot t = 0$

14. $n \cdot 5 = 15$

15. $22 \cdot 3 \cdot x = 594$

16. $t \cdot 0 = 25$

Objective 3 Solve applications involving division.
For extra help, see Example 8 on page 46 of your text and the Section 1.4 lecture video.
Solve.

17. If the area of a table top is 8 square feet and the width is 2 feet, find the length. Use the formula for the area of rectangle.

17. Five friends go out for dinner and the bill comes to $186.25. If they want to split the bill evenly, how much should they each pay?

Chapter 1 WHOLE NUMBERS

1.5 Order of Operations; Mean, Median and Mode

Learning Objectives
1   Simplify numerical expressions by following the order of operations agreement.
2   Find the mean, median, and mode of a list of values.

Key Terms

Use the vocabulary terms listed below to complete each statement in Exercises 1–4.

average　　　　**mean**　　　**median**　　　　**mode**

1. The _____ of a set of values is found by adding the values and dividing by the number of values.

2. Another word used for the mean of a set of values is _____.

3. The _____ of a set of values is the value that occurs most often.

4. In order to find the _____ of a set of values the first step is to arrange the values in order from least to greatest.

GUIDED EXAMPLES AND PRACTICE

Objective 1 Simplify numerical expressions by following the order of operations agreement.

Review these examples for Objective 1:

1. Simplify. $20 - 2^3 \cdot 1^2 + 13$

$20 - 2^3 \cdot 1^2 + 13$

$= 20 - 8 \cdot 1 + 13$ Evaluate the exponential form.

$$2^3 = 8, 1^2 = 1$$

$= 20 - 8 + 13$ Multiply. $8 \cdot 1 = 8$

$= 12 + 13$ Subtract. $20 - 8 = 12$

$= 25$ Add. $12 + 13 = 25$

2. Simplify. $3^4 \div (11 - 2) + 3(6 + 2) - (7 - 5)$

Practice these exercises:

1. Simplify.
$9 \cdot 2 + 6^2 \div 3 \cdot 2 - 5$

2. Simplify.
$\left[16 - 3(7 - 2) \right] + 3^3 - 4$

$$3^4 \div (11-2) + 3(6+2) - (7-5)$$

$= 3^4 \div (9) + 3(8) - (2)$ Perform operations in parentheses.

$= 81 \div 9 + 3(8) - 2$ Evaluate the exponential form. $3^4 = 81$

$= 9 + 24 - 2$ Multiply and divide from left to right. $81 \div 9 = 9$ and $3(8) = 24$

$= 31$ Add and subtract from left to right. $9 + 24 = 33$ and $33 - 2 = 31$

Objective 2 Find the mean, median, and mode of a list of values.

Review this example for Objective 2:

3. Find the mean, median and mode. 55, 58, 62, 69

$$\frac{55 + 58 + 62 + 69}{4} = 61 \leftarrow \text{Mean}$$

$54, 58, \quad 62, 69$

       $\uparrow$

     median

$$\text{median} = \text{mean of 58 and 62} = \frac{58 + 62}{2} = 60$$

Since no value is repeated, the data has no mode.

Practice this exercise:

3. Find the mean, median and mode.
54, 26, 49, 54, 63, 78

ADDITIONAL EXERCISES
Objective 1 Simplify numerical expressions by following the order of operations agreement.

For extra help, see Examples 1–6 on pages 50–52 of your text and the Section 1.5 lecture video.

Simplify.

1. $12 - 5 \cdot 2$

2. $16 \div 4 + 5 - 2^2$

3. $(12 - 2)\left[8 + 2^5 - 11 + 7(6 - 1)\right]$

4. $(12 - 7)2 + 6^2 - 2 \cdot 5$

5. $4^3 - 3 \cdot 2^2 \div 2 + 6(12 - 9)$

6. $32 \div 8 \cdot 2 + (21 \div 3) - 2^3$

7. $4 \times 3 + \left\{7 + 3\left[25 - 4(3 + 1)\right]\right\}$

8. $\dfrac{16 - 3^2(4 - 3)^2}{14 - 5 - 2}$

9. $4^2\left\{6\left[7 - 2(16 - 2^4) + 25\right]\right\} - 3^3 + 5(18 - 15)$

10. $\dfrac{16 - 2^4 + 3(8 - 5)^2}{(26 \div 2) + 7(2)}$

Objective 2 Find the mean, median, and mode of a list of values.

For extra help, see Example 7 on page 54 of your text and the Section 1.5 lecture video.

Find the arithmetic mean, median, and mode.

11. 80, 73, 78, 80, 85, 78

12. 125, 187, 168, 187, 107, 156

13. 201, 507, 628, 578, 546, 870

14. 313, 487, 951, 574, 278, 487

Chapter 1 WHOLE NUMBERS

1.6 More with Formulas

Learning Objectives
1   Use the formula $P = 2l + 2w$ to find the perimeter of a rectangle.
2   Use the formula $A = bh$ to find the area of a parallelogram.
3   Use the formula $V = lwh$ to find the volume of a box.
4   Solve for an unknown number in a formula.
5   Use a problem-solving process to solve problems requiring more than one formula.

Key Terms
Use the vocabulary terms listed below to complete each statement in Exercises 1–5.

parallel lines   parallelogram   right angle   cubic unit   volume

1.   Two lines that meet at a 90 degree angle form a _____.

2.   _____ is a measure of the amount of space inside a three-dimensional object.

3.   Two straight lines that never intersect are called _____.

4.   A cube that measures 1 unit by 1 unit by 1 unit has a volume of 1 _____.

5.   To find the area of a _____, multiply the measure of the base by the measure of the height.

GUIDED EXAMPLES AND PRACTICE

Objective 1   Use the formula $P = 2l + 2w$ to find the perimeter of a rectangle.

Review this example for Objective 1:
1.   Find the perimeter.

15 cm

7 cm

Practice this exercise:
1.   Find the perimeter.

21 in.

3 in.

$P = 2l + 2w$

$P = 2(15) + 2(7)$ Replace l with 15 and w with 7.

$P = 30 + 14$ Multiply.

$P = 44$ Add.

The perimeter is 44 cm.

Objective 2 Use the formula $A = bh$ to find the area of a parallelogram.

Review this example for Objective 2:

2. Find the area.

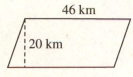

46 km

20 km

$A = bh$

$A = (46)(20)$ In $A = bh$ replace b with 46 and h with 20.

$A = 920$ Multiply.

The area is 920 km^2.

Practice this exercise:

2. Find the area.

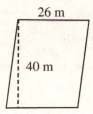

26 m

40 m

Objective 3 Use the formula $V = lwh$ to find the volume of a box.

Review this example for Objective 3:

3. Find the volume.

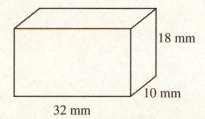

18 mm

10 mm

32 mm

$V = lwh$

$V = (32)(10)(18)$ Replace l with 32, w with 10, and h with 18.

$V = 5760$ Multiply.

The volume is 5760 mm^3.

Practice this exercise:

3. Find the volume.

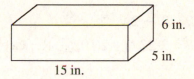

6 in.

5 in.

15 in.

Objective 4 Solve for an unknown number in a formula.

Review this example for Objective 4:

4. Solve. The area of a parallelogram is 96 square centimeters. The base of the figure is 16 centimeters. What is the height?

Use the formula $A = bh$.

$A = bh$

$96 = 16h$ Replace A with 96 and
                  b with 16.

$96 \div 16 = h$ Write a related division
                  statement.

$6 = h$ Divide.

The height is 6 cm.

Practice this exercise:

4. Solve. The area of a rectangular room is 192 square feet. The width is 16 feet. What is the length?

Objective 5 Use a problem-solving process to solve problems requiring more than one formula.

Review this example for Objective 5:

5. Solve. A security fence is to be built around a 237-meter by 102-meter field. What is the perimeter of the field? Fence wire will cost $3 per meter. What will the fence cost?

Use $P = 2l + 2w$ to find the perimeter.

$P = 2l + 2w$

$P = 2(237) + 2(102)$ Replace l with 237
                       and w with 102.

$P = 474 + 204$ Multiply.

$P = 678$ Add.

The perimeter is 678 m.

Multiply the length of the perimeter by the cost per meter to find the total cost.

$678 \times 3 = 2034$

The fence will cost $2034.

Practice this exercise:

5. Solve. Dean's backyard is a rectangle 113 feet by 28 feet. Last year he put a 30-foot by 14-foot brick patio in the yard. How much of the yard is left to mow?

Name: Date:
Instructor: Section:

ADDITIONAL EXERCISES

Objective 1 Use the formula $P = 2l + 2w$ to find the perimeter of a rectangle.

For extra help, see Example 1 on page 58 of your text and the Section 1.6 lecture video.
Solve.

1. A picket fence is to be built around a
    218-foot by 94-foot yard. What is the
    perimeter of the yard?

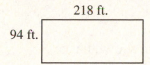

2. A rectangle has a length of
    72 cm and a width of 31 cm.
    Find the perimeter.

Objective 2 Use the formula $A = bh$ to find the area of a parallelogram.

For extra help, see Example 2 on page 60 of your text and the Section 1.6 lecture video.
Find the area.

3.

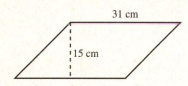

4.

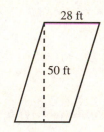

Objective 3 Use the formula $V = lwh$ to find the volume of a box.

For extra help, see Example 3 on page 61 of your text and the Section 1.6 lecture video.
Find the volume.

5.

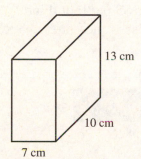

6.

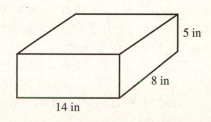

Objective 4 Solve for an unknown number in a formula.

For extra help, see Examples 4–5 on pages 61–62 of your text and the Section 1.6 lecture video.

Solve.

7. The area of a parallelogram is 294 square centimeters. The height of the figure is 14 centimeters. What is the base?

8. The area of a rectangular room is 405 square feet. The length is 27 feet. What is the width?

9. The figure shown has a volume of 144 cubic centimeters. Find the height of the figure.

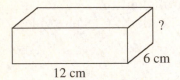

10. The figure shown has a volume of 8 cubic centimeters. Find the width of the figure.

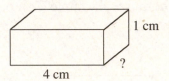

Objective 5 Use a problem-solving process to solve problems requiring more than one formula.

For extra help, see Examples 6–9 on pages 63–67 of your text and the Section 1.6 lecture video.

Solve.

11. A room is 13 feet by 21 feet. It needs to be painted. The ceiling is 8 feet above the floor. There are two windows in the room; each one is 3 feet by 4 feet. The door is 3 feet by 7 feet.

    a. Find the area of the walls.

    b. One gallon of paint covers 93 square feet. How many gallons are needed to paint the walls?

    c. At $23 per gallon, what is the cost to paint the walls?

12. A rain gutter is to be installed around the house shown. The gutter costs $5 per foot. Find the total cost of the gutter.

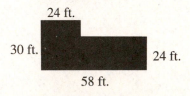

Chapter 2 INTEGERS

2.1 Introduction to Integers

Learning Objectives
1.   Identify integers.
2.   Interpret number lines and graphs with integers.
3.   Use < or > to write a true statement.
4.   Find the absolute value.

Key Terms

Use the vocabulary terms listed below to complete each statement in Exercises 1–5.

      integers   **absolute value**   **additive inverse**   **positive**   **negative**

1.   On the number line, the distance a number is away from 0 is the number's
   _____.

2.   Distance can never be a _____ value.

3.   The _____ are the whole numbers and all of their additive inverses.

4.   The _____ of a positive number is negative.

5.   The absolute value of every nonzero number is _____.

GUIDED EXAMPLES AND PRACTICE

Objective 1 Identify integers.

Review these examples for Objective 1:

1. Express the amount as a positive or negative integer.
   A woman owes her friend $116.

   Money owed is debt, which is represented by a negative number. −116

2. A geologist visits a site 2953 feet above sea level. Express this amount as a positive or negative integer.

   Since 2953 feet is above sea level, it is represented by a positive number. +2953

Practice these exercises:

1. The record low temperature for a date is 21 degrees below zero. Express this amount as a positive or negative integer.

2. During a year, a person saves $1350. Express this amount as a positive or negative integer.

Objective 2 Interpret number lines and graphs with integers.

Review these examples for Objective 2:	Practice these exercises:

3. Graph −4 on a number line.

Draw a dot on the mark of −4.

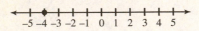

3. Graph 3 on a number line.

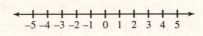

For exercise 4, use the following graph.

Maximum and Minimum Normal Temperatures in January (degrees Fahrenheit)

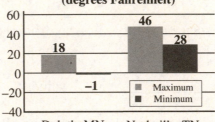

Duluth, MN   Nashville, TN

4. What is the minimum normal January temperature in Nashville, Tennessee?

28°F

4. What is the maximum normal January temperature in Duluth, Minnesota?

Objective 3 Use < or > to write a true statement.

Review this example for Objective 3:	Practice this exercise:

5. Use either < or > for ☐ to form a true sentence.

−3 ☐ 4

Since −3 is to the left of 4 on the number line, we have −3 < 4.

5. Use either < or > for ☐ to form a true sentence.

−10 ☐ −6

Name: Date:
Instructor: Section:

Objective 4 Find the absolute value.

Review these examples for Objective 4:

6. Find the absolute value of −21.

    Because −21 is 21 steps from 0 on a number line, its absolute value is 21.

7. Find the absolute value of 10.

    Because 10 is 10 steps from 0 on a number line, its absolute value is 10.

8. Find $|109|$.

    Because 109 is 109 steps from 0 on a number line, its absolute value is 109.

9. Find $|-31|$.

    Because −31 is 31 steps from 0 on a number line, its absolute value is 31.

Practice these exercises:

6. Find the absolute value of −59.

7. Find the absolute value of 33.

8. Find $|46|$.

9. Find $|-17|$.

ADDITIONAL EXERCISES
Objective 1 Identify integers.

For extra help, see Example 1 on page 89 of your text and the Section 2.1 lecture video.
Tell which integers correspond to each situation.

1. Express the amount as a positive or negative integer.

    Steven owes $416 on a car loan.

2. Express the amount as a positive or negative integer.

    Stan receives a paycheck of $562.

3. Express the amount as a positive or negative integer.

    The owner of a diner lowered her prices on sandwiches by $1.50.

4. Express the amount as a positive or negative integer.

    The population of Hudson City decreased by 3142 people.

Name: _____ Date: _____
Instructor: _____ Section: _____

Objective 2 Interpret number lines and graphs with integers.
For extra help, see Examples 3–4 on page 90 of your text and the Section 2.1 lecture video.
Graph each integer on a number line.

5. 6

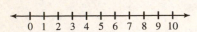

6. 7

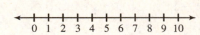

7. −7

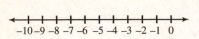

8. −10

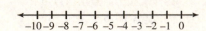

Objective 3   Use < or > to write a true statement.
For extra help, see Example 4 on page 91 of your text and the Section 2.1 lecture video.

Use either < or > for ☐ *to form a true sentence.*

9. 3 ☐ −7

10. −2 ☐ 0

11. −72 ☐ −56

12. −8 ☐ 3

Objective 4   Find the absolute value.
For extra help, see Examples 5–6 on page 91–92 of your text and the Section 2.1 lecture video.
Find the absolute value.

13. |−26|

14. |41|

15. |−175|

16. |0|

Chapter 2 INTEGERS

2.2 Adding Integers

Learning Objectives
1   Add integers with like signs.
2   Add integers with different signs.
3   Add integers.
4   Find the additive inverse
5   Solve applications involving addition of integers.

GUIDED EXAMPLES AND PRACTICE

Objective 1 Add integers with like signs.

Review this example for Objective 1:
1. Add without using a number line:

$28 + 34$ *Think*: Add the abosolute

                values: $28 + 34 = 62$.

                Make the answer positive, 62.

Practice this exercise:
1. Add without using a number line:

$-18 + (-14)$

Objective 2 Add integers with different signs.

Review this example for Objective 2:
2. Add without using a number line:

$48 + (-83)$ *Think*: The absolute values

                are 48 and 83. The difference is

                35. Since the negative number

                has the larger absolute value,

                the answer is *negative*, -35.

Practice this exercise:
2. Add without using a number line:

$62 + (-72)$

Objective 3 Add integers.

Review these examples for Objective 3:
3. Explain the situation in terms of debts and credits, then add.

$52 + (-12)$

A credit with a debt, so subtract and because the credit, 52, is more than the debt, the result is a *credit/positive*; 40.

Practice these exercises:
3. Explain the situation in terms of debts and credits, then add.

$41 + (-63)$

4. Simplify $18+(-3)+(-7)+15$.

$18+(-3)+(-7)+15$

$15+15+(-7)$

$30+(-7)$

23

4. Simplify $-72+(-1)+36+48$.

Objective 4 Find the additive inverse.

Review this example for Objective 4:

5. Find the additive inverse of -28.

-28 and 28 are additive inverses
because $-28 + 28 = 0$. Also, they
have the same absolute value but
opposite signs.

Practice this exercise:

5. Find the additive inverse of 83.

Objective 5 Solve applications involving addition of integers.

Review this example for Objective 5:

6. The barometric pressure at a certain
city dropped 8 millibars (mb); then
it rose 3 mb. After that, it dropped
6 mb and then it rose
6 mb. What was the total change in
pressure?

Add the changes in pressure.
$-8+3+(-6)+6$

$-5+(-6)+6$

$-11+6$

-5

The total change in pressure
is -5 mb.
The pressure dropped 5 mb.

Practice this exercise:

6. The rim of a canyon is at an
altitude of 76 m. On a hike down
into the canyon, a party of hikers
stops for a rest 108 m below the
rim. Then they descend another
60 m. What is their new altitude?

Name: Date:
Instructor: Section:

ADDITIONAL EXERCISES
Objective 1   Add integers with like signs.
For extra help, see Example 1 on page 96 of your text and the Section 2.2 lecture video.
Add without using a number line.

1. $18+9$

2. $-7+(-6)$

3. $-9+(-2)$

4. $13+17$

5. $-8+(-16)+(-31)$

6. $-12+(-61)+(-21)$

Objective 2   Add integers with different signs.
For extra help, see Example 2 on page 97 of your text and the Section 2.2 lecture video.
Add.

7. $8+(-16)$

8. $-4+4$

9. $-19+7$

10. $0+(-11)$

Objective 3   Add integers.
For extra help, see Examples 3–4 on pages 97–98 of your text and the Section 2.2 lecture video.
Add.

11. Explain the situation in terms of debts and credits, then add.
    $31+(-43)$

12. Explain the situation in terms of debts and credits, then add.
    $-11+67$

13. $2+(-4)+13+(-7)$

14. $-6+1+(-2)+5$

Objective 4   Find the additive inverse.
For extra help, see Example 5 on page 99 of your text and the Section 2.2 lecture video.
Find the additive inverse.

15. −74   16. 8

17. 4   18. −513

Objective 5   Solve applications involving addition of integers.
For extra help, see Examples 6–8 on pages 100–101 of your text and the Section 2.2 lecture video.

19. Kyle's credit card bill is $480. Kyle sends a check to the credit card company for $53, charges another $171 in merchandise, and then pays off another $240 of the bill. How much does Kyle owe the company?

20. An airplane climbs to 35,000 feet, and then descends 2340 feet. What is its new altitude?

21. A submarine uses compressed air to launch a rocket from 35 feet below the surface of the ocean. The rocket rises 100 feet before it fires its own engine. At what height does the rocket fire its engine?

22. Charlie finds a $5 bill in the street and uses it to buy a $1 chocolate bar. How much money does he have left?

Chapter 2 INTEGERS

2.3 Subtracting Integers and Solving Equations

Learning Objectives
1   Write subtraction statements as equivalent addition statements.
2   Solve equations containing an unknown addend.
3   Solve applications involving subtraction of integers.

Key Terms

Use the vocabulary terms listed below to complete each statement in Exercises 1–4.

> **net**   **cost**   **revenue**   **profit**   **loss**

1. _____ is the amount spent.

2. After the amount spent is subtracted from the amount made, the amount left is the
    _____.

3. The result when revenue is less than cost is a _____.

4. To obtain a profit, _____ must be greater than cost.

GUIDED EXAMPLES AND PRACTICE

Objective 1 Write subtraction statements as equivalent addition statements.

Review these examples for Objective 1:

1. Subtract: $6 - 13$
    Change the operation symbol
    from − to +.
    Change the second number to its
    additive inverse.

    $6 + (-13)$

    -7

2. Subtract: $19 - (-21)$
    Change the operation symbol
    from − to +.
    Change the second number to its
    additive inverse.

    $19 + 21$

    40

Practice these exercises:

1. Subtract: $15 - 37$

2. Subtract: $-14 - (-16)$

Objective 2 Solve equations containing an unknown addend.

Review these examples for Objective 2:	Practice these exercises:

Review these examples for Objective 2:

3. Solve and check: $15 + p = 12$

Write a related subtraction equation in which the known addend is subtracted from the sum.

$p = 12 - 15$

$p = 12 + (-15)$

$p = -3$

Check. Verify that the equation is true.

$15 + (-3) \overset{?}{=} 12$

$12 = 12$

4. Solve and check: $r + (-2) = 7$

Write a related subtraction equation in which the known addend is subtracted from the sum.

$r = 7 - (-2)$

$r = 7 + 2$

$r = 9$

Check. Verify that the equation is true.

$9 + (-2) \overset{?}{=} 7$

$7 = 7$

Practice these exercises:

3. Solve and check:

$27 + x = 11$

4. Solve and check:

$d + (-17) = 13$

Objective 3 Solve applications involving subtraction of integers

Review this example for Objective 3:

5. One day the temperature dropped from $-1°\,F$ to $-14°\,F$. How many degrees did it drop?

Calculate the difference.

$-14 - (-1)$

$-14 + 1$

-13

The change in temperature was -13 degrees. The temperature dropped 13 degrees.

Practice this exercise:

5. A submarine at a depth of 1881 ft ascends to a depth of 838 ft. How far did the submarine ascend?

ADDITIONAL EXERCISES
Objective 1 Write subtraction statements as equivalent addition statements.
For extra help, see Examples 1–2 on pages 106–108 of your text and the Section 2.3 lecture video.
Write as an equivalent addition, then evaluate.

1. $8 - 4$

2. $-15 - 31$

3. $21 - 16$

4. $-316 - 84$

Objective 2 Solve equations containing an unknown addend.
For extra help, see Example 3 on page 108 of your text and the Section 2.3 lecture video.
Solve and check.

5. $x + (-3) = 9$

6. $-1 + m = -17$

7. $f - 2 = 81$

8. $r + (-3) = -2$

Objective 3   Solve applications involving subtraction of integers.
For extra help, see Examples 4–6 on pages 109–110 of your text and the Section 2.3 lecture video.
Find the absolute value.

9. During one year, a business had total revenues of \$2,620,760 and total costs of \$1,620,260. Calculate the net and determine if it was a profit or a loss.

10. Hikers descended a 2500 foot mountain, and set up camp 400 feet from the bottom of the mountain. Find how far the hikers descended from the top of the mountain to the camp.

Chapter 2 INTEGERS

2.4 Multiplying and Dividing Integers; Exponents; Square Roots; Solving Equations

Learning Objectives
1   Multiply integers.
2   Evaluate numbers in exponential form.
3   Divide integers.
4   Solve equations containing an unknown factor.
5   Evaluate square roots.
6   Solve applications involving multiplication or division of integers.

Key Terms
Use the vocabulary terms listed below to complete each statement in Exercises 1–6.

negative   positive   same   different   even   odd

1.   When evaluating an exponential form that has a negative base, if the exponent is _____ the product is positive.

2.   When multiplying two numbers that have different signs, the product is _____ .

3.   When multiplying signed numbers, if there is an even number of negative numbers the product is _____ .

4.   The product of two numbers with the _____ sign is always positive.

5.   The principal square root is the _____ square root of a number.

6.   The square root of a _____ number is not an integer.

GUIDED EXAMPLES AND PRACTICE

Objective 1 Multiply integers.

Review these examples for Objective 1:
1.   Multiply $-6 \cdot 5$
    The signs are different, so the answer is negative.
    $-6 \cdot 5 = -30$

Practice these exercises:
1.   Multiply $5 \cdot (-8)$

2. Multiply: $-12 \cdot (-11)$

The signs are the same, so the answer is positive.

$$-12 \cdot (-11) = 132$$

2. Multiply: $-5 \cdot (-10)$

3. Multiply: $-1(-2)(-1)$

Multiply from left to right

$$-1(-2)(-1)$$

$$2(-1)$$

$$-2$$

3. Multiply: $-3(-5)(-1)$

Objective 2 Evaluate numbers in exponential form.

Review these examples for Objective 2:

4. Evaluate: $(-6)^2$

The exponent is even. The product is positive.

$$(-6)^2 = (-6)(-6) = 36$$

Practice these exercises:

4. Evaluate: $(-2)^3$

5. Evaluate: -2^6

The minus sign is an understood -1 multiplying 2^6.

$$-2^6$$

$$= -1 \cdot 2^6$$

$$= -1 \cdot 2 \cdot 2 \cdot 2 \cdot 2 \cdot 2 \cdot 2$$

$$= -1 \cdot 64$$

$$= -64$$

5. Evaluate: -5^2

Objective 3 Divide integers.

Review this example for Objective 3:

6. Divide: $14 \div (-2)$

The signs are different, so the answer is negative.

$$14 \div (-2) = -7$$

Practice this exercise:

6. Divide: $-21 \div (-3)$

Objective 4 Solve equations containing an unknown factor.

Review this example for Objective 4:	**Practice this exercise:**

7. Solve and check: $-11a = 77$

Write a related division equation in which the product is divided by the known factor.

$$a = \frac{77}{-11} = -7$$

Check. Verify that the equation is true.

$$-11(-7) \overset{?}{=} 77$$
$$77 = 77$$

7. Solve and check: $-1f = -6$

Objective 5 Evaluate square roots.

Review these examples for Objective 5:	**Practice these exercises:**

8. Find all the square roots of 9.

3 and -3, because $3^2 = 9$ and $(-3)^2 = 9$.

8. Find all the square roots of 169.

9. Simplify: $\sqrt{16}$

The positive square root of 16 is 4.

9. Simplify: $-\sqrt{121}$

Objective 6 Solve applications involving multiplication or division of integers.

Review this example for Objective 6:	**Practice this exercise:**

10. An estate of $4277 was divided equally among seven family members. Find the amount received by each family member.

Divide the amount of money by the number of family members.

$$4277 \div 7 = 611$$

Each family member receives $611.

10. Kevin has three children in college. For each child, Kevin pays tuition of $13,500 per year. What is the total tuition   that Kevin pays per year?

Name: Date:
Instructor: Section:

ADDITIONAL EXERCISES
Objective 1   Multiply integers.
For extra help, see Examples 1–4 on pages 114–116 of your text and the Section 2.4 lecture video.
Multiply.

1. $8 \cdot (-4)$

2. $-7 \cdot (-5)$

3. $-13 \cdot 0$

4. $-9 \cdot (-2)$

5. $(-5)(-6)(-2)(4)$

6. $-3(-18)0 \cdot 15$

Objective 2   Evaluate numbers in exponential form.
For extra help, see Examples 5–6 on page 117 of your text and the Section 2.4 lecture video.
Evaluate.

7. $(-2)^4$

8. $(-2)^5$

9. -5^2

10. $(-11)^2$

Objective 3   Divide integers.
For extra help, see Examples 7–8 on pages 118–119 of your text and the Section 2.4 lecture video.
Divide.

11. $18 \div (-2)$

12. $-51 \div (-3)$

13. $16 \div 0$

14. $-21 \div (-7)$

Objective 4   Solve equations containing an unknown factor.
For extra help, see Example 8 on pages 118–119 of your text and the Section 2.4 lecture video.
Solve and check.

15. $-2y = -6$

16. $f \cdot 7 = -63$

17. $4r = -72$

18. $(-5)(2)m = 130$

Objective 5   Evaluate square roots.
For extra help, see Examples 9–10 on pages 119–120 of your text and the Section 2.4 lecture video.
Find all the square roots of each number.

19. 144

20. 64

Simplify.

21. $\sqrt{25}$

22. $-\sqrt{49}$

Objective 6 Solve applications involving multiplication or division of integers.
For extra help, see Example 11 on page 121 of your text and the Section 2.4 lecture video.
Solve.

23. A family of four bought tickets to a baseball game. If the tickets were $34 each, how much did it cost for the family to go to the game?

Chapter 2 INTEGERS

2.5 Order of Operations

Learning Objectives
1   Simplify numerical expressions by following the order of operations agreement.

GUIDED EXAMPLES AND PRACTICE

Objective 1 Simplify numerical expressions by following the order of operations agreement.

Review these examples for Objective 1:

1. Simplify:

$-16 \div 4 \cdot 2$

$= -4 \cdot 2$ Divide $-16 \div 4 = -4$

$= -8$ Multiply $-4 \cdot 2 = -8$

2. Simplify:

$\sqrt{16} + \sqrt{9}$

$= 4 + 3$ Evaluate the square roots.

$\sqrt{16} = 4, \sqrt{9} = 3$

$= 7$ add $4 + 3 = 7$

3. Simplify:

$-24 \div 4 \left[5 - (6+8) \div 7 \right] + |-7|$

$-24 \div 4 \left[5 - (14) \div 7 \right] + |-7|$ Add in the parentheses

$6 + 8 = 14$

$-24 \div 4 \left[5 - 2 \right] + |-7|$ Divide $14 \div 7 = 2$

$-24 \div 4 \left[3 \right] + |-7|$ Subtract $5 - 2 = 3$

$-24 \div 4 \left[3 \right] + 7$ Find the absolute value

of -7

$-6 \left[3 \right] + 7$ Divide $-24 \div 4 = -6$

$-18 + 7$ Multiply $-6 \cdot 3 = -18$

-11 Add $-18 + 7$

Practice these exercises:

1. Simplify: $18 - 2 \cdot 16$

2. Simplify: $\sqrt{16+9}$

3. Simplify:

$10 - 8 \left[-15 - 2 \div (-2) \right] + 3^3$

4. Simplify:

$2|7 - 2 \cdot 4| + 18 - 2^3$

$2|7 - 8| + 18 - 2^3$ Multiply in the absolute value

symbols $2 \cdot 4 = 8$

$2|-1| + 18 - 2^3$ Subtract in the absolute value

symbols $7 - 8 = -1$

$2(1) + 18 - 2^3$ Find the absolute value of -1

$2(1) + 18 - 8$ Evaluate the exponential form

$2^3 = 8$

$2 + 18 - 8$ Multiply $2 \cdot 1 = 2$

$20 - 8$ Add $2 + 18 = 20$

12 Subtract $20 - 8 = 12$

4. Simplify:

$-|-40 \div (-2) + 6| + 7\sqrt{4}$

5. Simplify:

$$\frac{14 - 6(4)}{(8-4)+6}$$

$$= \frac{14 - 24}{(8-4)+6}$$ Multiply in the numerator $6 \cdot 4 = 24$

$$= \frac{14 - 24}{(4)+6}$$ Subtract in the denominator $8 - 4 = 4$

$$= \frac{-10}{4+6}$$ Subtract in the numerator $14 - 24 = -10$

$$= \frac{-10}{10}$$ Add in the numerator $4 + 6 = 10$

$$= -1$$ Divide $-10 \div 10 = -1$

5. Simplify:

$$\frac{-3^2 + 2(7-12)^2 + 3}{16 - 20 + (8-4)^2 - 1}$$

Name:
Instructor:

Date:
Section:

ADDITIONAL EXERCISES
Objective 1 Simplify numerical expressions by following the order of operations agreement.
For extra help, see Examples 1–10 on pages 124–127 of your text and the Section 2.5 lecture video.
Simplify.

1. $-32 \div 8 \cdot 3$

2. $(9+5) \div 7 \cdot 3$

3. $14 - 2 \cdot 8$

4. $63 \div (-9) \cdot (-2)$

5. $-2^3 (-14 + 17)^2 + \sqrt{36}$

6. $-(12-1)^2 + \sqrt{49}$

7. $(-3)^2 + 2(14 - 20) + 6^2$

8. $-5^2 + 12 \div 6 + 6^2$

9. $\sqrt{100} - \sqrt{36}$

10. $\sqrt{100 - 36}$

11. $2|7 - 2 \cdot 4| + 18 - 2^3$

12. $-|-40 \div (-2) + 6| + 7\sqrt{4}$

13. $-6\sqrt{25} + 15 \div (-5) - 7(-2)$

14. $2\sqrt{50 - 5 \cdot 5} + 7(-3)^2 - 2^5$

15. $\dfrac{-|15(16 - 12)| - 6(-10)}{5\{7 - 3[2(6 - 10) + 8]\} - 2^4}$

16. $\dfrac{-96 + 5[6(4 - 2)^2 + 7] - 12(-7)}{-6|16 \div [2(-4)]|^3 + 48}$

Chapter 2 INTEGERS

2.6 Additional Applications and Problem Solving

Learning Objectives
1   Solve problems involving net.
2   Solve problems involving voltage.
3   Solve problems involving average rate.

GUIDED EXAMPLES AND PRACTICE

Objective 1   Solve problems involving net.

Review this example for Objective 1

1. Tomas put $1000 down when he bought a new car. He made 36 payments of $326 and spent $750 in maintenance and repairs. Three years after paying off the car he sold it for $5200. What was his net? Was it a profit or loss?

Use the formula $N = R - C$ to calculate the net.
First find the total cost for the car.
Cost = amount down + total of all payments+ maintenance cost.

$$Cost = 1000 + 36(326) + 750$$

$$Cost = 1000 + 11,736 + 750$$

$$Cost = 13,486$$

Now use $N = R - C$ to calculate the net.

$$N = 5200 - 13486$$

$$N = -8286$$

Tomas' net is $-\$8286$, which is a loss of $\$8286$.

Practice this exercise:

1. Tanya takes out a loan to buy a fixer-upper house. She then spends $4900 in repairs and improvements. She sells the house for $80,880. The payoff amount for the loan that she took out to buy the house is $70,974. What is her net? Is it a profit or loss?

Objective 2 Solve problems involving voltage.

Review this example for Objective 2

2. The voltage in an electrical circuit measures -252 volts. If the resistance is 7 ohms, find the current.

   Use the formula $V = ir$, where V is the voltage, and r is the resistance.

   $$V = ir$$
   $$-252 = i \cdot 7$$
   $$\frac{-252}{7} = i$$
   $$-36 = i$$

   The current is -36 amps.

Practice this exercise:

2. An electrical circuit has a resistance of 45 ohms and a current of -11 amps. Find the voltage.

Objective 3 Solve problems involving average rate.

Review this example for Objective 3

3. A car traveled 268 miles in 4 hours. How far did it travel in 1 hour?

   Use the formula $d = rt$, where d is the distance, and t is the time.

   $$d = rt$$
   $$268 = r \cdot 4$$
   $$\frac{268}{4} = r$$
   $$67 = r$$

   The car traveled 67 miles in 1 hour.

Practice this exercise:

3. A plane flies 151 kilometers per hour for 3 hours. How far does it travel?

ADDITIONAL EXERCISES

Objective 1 Solve problems involving net.

For extra help, see Example 1 on page 131 of your text and the Section 2.6 lecture video.
Solve.

1. In 1995, Priscilla bought a collection of four ceramic village pieces. Two of the buildings cost $125 each and the other two cost $175 each. In 2007 she sold the entire collection in an auction for $725. What was her net? Was it a profit or loss?

2. Jason bought 60 shares of stock at a price of $72 per share. One day, he sold 20 shares for $66 per share. A week later, he sold the rest of his shares for $58 per share. What was his net? Was it a profit or loss?

Objective 2 Solve problems involving voltage.

For extra help, see Example 2 on page 132 of your text and the Section 2.6 lecture video.
Solve.

3. The voltage in an electrical circuit measures -512 volts. If the resistance is 8 ohms, find the current.

4. An electrical circuit has a resistance of 18 ohms and a current of -13 amps. Find the voltage.

Objective 3 Solve problems involving average rate.

For extra help, see Examples 3–4 on page 133 of your text and the Section 2.6 lecture video.
Solve.

5. On a scenic drive, Ava left at 9 A.M. and drove 75 miles, then took a 15-minute break. She then traveled another 50 miles and stopped for 1 hour to eat lunch. Finally she traveled 122 miles and arrived at her destination at 3 P.M. What was her average speed?

6. A small airplane flew 648 miles in 8 hours. How far did it fly in 1 hour?

Chapter 3 EXPRESSIONS AND POLYNOMIALS

3.1 Translating and Evaluating Expressions

Learning Objectives
1 Differentiate between an expression and an equation.
2 Translate word phrases to expressions
3 Evaluate expressions.

Key Terms
Use the vocabulary terms listed below to complete each statement in Exercises 1--5.

equation   expression   evaluate   undefined   indeterminate

1. To _____ an expression, replace the variables with the corresponding given values.

2. An _____ is a mathematical relationship that has an equal sign.

3. The quotient $\dfrac{0}{0}$ is said to be _____.

4. An _____ may be a combination of constants, variables, and arithmetic symbols.

5. For any number $n \neq 0$, the quotient $\dfrac{n}{0}$ is _____.

GUIDED EXAMPLES AND PRACTICE

Objective 1 Differentiate between an expression and an equation.

Review these examples for Objective 1:	Practice these exercises:
1. Determine whether the following is an expression or an equation. $3x + 5(x-1)$ $3x + 5(x-1)$ is an expression because it does not contain an equals sign.	1. Determine whether the following is an expression or an equation. $1 + \sqrt{7} = 8x$
2. Determine whether the following is an expression or an equation. $4x + 13 = 6(3-x)$ $4x + 13 = 6(3-x)$ is an equation because it contains an equals sign.	2. Determine whether the following is an expression or an equation. $\dfrac{2x^2 + 1}{x - 7}$

Objective 2 Translate word phrases to expressions.

Review these examples for Objective 2:

3. Translate to a variable expression.
    Eight subtracted from the product of five and x.

    The product of five and x translates to $5x$.

    Eight subtracted from $5x$ translates to $5x - 8$.

4. Translate the words to a variable expression.
    The product of nine and the sum of some number and two.

    "The sum of some number and two" translates to $n + 2$ when n is the unknown number.
    Nine times this sum translates to $9(n + 2)$

Practice these exercises:

3. Translate to a variable expression.
    The product of five and the square of n.

4. Translate the words to a variable expression.
    The quotient of the difference of x and y and the sum of x and four.

Objective 3 Evaluate expressions.

Review these examples for Objective 3:

5. Evaluate $1 + y;\ y = 12$.

    $1 + (12)$ Replace y with 12 using parentheses

    13 Simplify by adding

6. Evaluate $x^2 - 6y + 9;\ x = -4,\ y = 3$.

    $(-4) - 6(3) + 9$ Replace x with -4 and y with -3.

    $-4 - 18 + 9$ Multiply 6 and 3.

    -13 Simplify by adding.

Practice these exercises:

5. Evaluate.
    $4x - 2y;\ x = 1,\ y = -7$

6. Evaluate.
    $\dfrac{-4mn + 2}{6n};\ m = -2,\ n = -1$

ADDITIONAL EXERCISES

Objective 1 Differentiate between an expression and an equation.

For extra help, see Example 1 on page 146 of your text and the Section 3.1 lecture video.

Determine whether each of the following is an expression or an equation.

1. $|14x - 9| = 6$

2. $\sqrt{3 - \dfrac{x}{c}}$

Objective 2 Translate word phrases to expressions.
For extra help, see Examples 2–3 on pages 148–149 of your text and the Section 3.1 lecture video.
Translate to a variable expression.

3. Negative three multiplies by the difference of some number and four.

4. Four minus the quotient of two and x.

Objective 3 Evaluate expressions.
For extra help, see Example 4 on page 149 of your text and the Section 3.1 lecture video.
Evaluate.

5. $8x - y$; $x = 5$, $y = 40$

6. $x^2 - 7y + 9$; $x = -1$, $y = -2$

7. $b^2 - 4ac$; $a = -3$, $b = 9$, $c = -4$

8. $2t + 1 + 4t - 5t$; $t = -1$

9. $-|-x - y|$; $x = -3$, $y = 18$

10. $\sqrt{c^2 - b^2}$; $c = 15$, $b = 9$

11. $\sqrt{m^2 - n}$; $m = -5$, $n = 9$

12. $\sqrt{\dfrac{7x - 3}{3x + 1}}$; $x = 1$

13. $\dfrac{-m + 26n}{s + 5}$; $m = 5$, $n = -5$, $s = 0$

14. $\dfrac{x + h}{h}$; $x = 2$, $h = 1$

Complete each table by evaluating the expression using the given values for the variable.

15.

x	$-3x + 5$
-2	
-1	
0	
1	
2	

16.

y	$|y^2 - 4|$
-4	
-2	
0	
2	
4	

Find all values for the variable that cause the expression to be undefined or indeterminate.

17. $\dfrac{x + 1}{x - 4}$

18. $\dfrac{12}{(x + 2)(x - 3)}$

Chapter 3 EXPRESSIONS AND POLYNOMIALS

3.2 Introduction to Polynomials; Combining Like Terms

Learning Objectives
1 Identify monomials.
2 Identify the coefficient and degree of a monomial.
3 Identify polynomials and their terms.
4 Identify the degree of a polynomial.
5 Write polynomials in descending order of degree.
6 Simplify polynomials in one variable by combining like terms.

Key Terms
Use the vocabulary terms listed below to complete each statement in Exercises 1–6.

monomial	polynomial	coefficient	degree	like
simplest	one variable	multivariable	binomial	trinomial
descending	ascending	greatest	unlike	least

1. The degree of a polynomial is the _____ of any of the degrees in the polynomial.

2. An expression is in _____ form if it is written with the fewest symbols possible.

3. The numerical factor in a term is its _____.

4. A polynomial with three variables is called a(n) _____ polynomial.

5. The _____ of a monomial is the sum of the exponents of all the variables in the monomial.

6. A constant term is a(n) _____.

GUIDED EXAMPLES AND PRACTICE

Objective 1 Identify monomials.

Review this example for Objective 1:

1. Determine whether the expression is a monomial. Explain.

   $6y^4$

   $6y^4$ is a monomial since it is a product of a constant, 6, and a variable, y, that has a whole number power, 4.

Practice this exercise:

1. Determine whether the expression is a monomial. Explain.

   $15m^3 - 12m^2 - 1$

Objective 2 Identify the coefficient and degree of a monomial.

Review this example for Objective 2:

2. Identify the coefficient and degree of each monomial.

   $7x^5$

   The coefficient is 7 because it is the numerical factor. The degree is 5 because it is the exponent of the variable x.

Practice this exercise:

2. Identify the coefficient and degree of each monomial.

   $-x$

Objective 3 Identify polynomials and their terms.

Review these examples for Objective 3:

3. Identify the terms in each polynomial and their coefficients.

   $15x^3 + 7x^2 + 4x - 21$

   First term: $15x^3$; Coefficient: 15
   Second term: $7x^2$; Coefficient: 7
   Third term: $4x$; Coefficient: 4
   Fourth term: -21; Coefficient: -21

3. Indicate whether the expression is a monomial, binomial, trinomial, or has no special name.

   $8m^2 - 5$
   $8m^2 - 5$ is a binomial because it has two terms, $8m^2$ and -5.

Practice these exercises:

3. Identify the terms in each polynomial and their coefficients.

   $3c^5 - 8c^3 - 18$

4. Indicate whether the expression is a monomial, binomial, trinomial, or has no special name.

   $21q^4 - 5q^2 + 7q + 11$

Objective 4 Identify the degree of a polynomial.

Review this example for Objective 4:

5. Identify the degree of the polynomial.
    $$5p^2 + 2p - 1$$

    The degree of the terms $5p^2$, $2p$, and -1 are 2, 1, and 0 respectively. The degree of $5p^2 + 2p - 1$ is 2 since it is the greatest degree of all the terms.

Practice this exercise:

5. Identify the degree of the polynomial.
    $$4d^5 - 6 + 7c^3 + 3c^2 - 2d^7$$

Objective 5 Write polynomials in descending order of degree.

Review this example for Objective 5:

6. Write each polynomial in descending order of degree.
    $$9a + 5a^4 - 21a^3 + 7 - 2a^2$$

    Move the term $5a^4$ to the front since it has the highest degree, 4. Next is $-21a^3$ with degree 3, $-2a^2$ with degree 2, $9a$ with degree 1. The constant term 7 is last since it has degree 0. The polynomial in descending order of degree is
    $$5a^4 - 21a^3 - 2a^2 + 9a + 7.$$

Practice this exercise:

6. Write each polynomial in descending order of degree.
    $$12z^2 + 14z^3 - 15 + z$$

Objective 6 Simplify polynomials in one variable by combining like terms.

Review these examples for Objective 6:

7. Determine whether the monomials are like terms.
    $$-4x^3 \text{ and } 2x^3$$

    $-4x^3$ and $2x^3$ are like terms since they have the same variable, x, raised to the same exponent, 3.

8. Combine like terms.
    $$-4z^5 + 7z^5$$

    Add the coefficients -4 and 7, and keep the variable z, and its exponent 5 the same. This gives the monomial
    $$(-4 + 7)z^3 = 3z^3.$$

Practice these exercises:

7. Determine whether the monomials are like terms.
    $$y^6 \text{ and } 6y$$

8. Combine like terms.
    $$9a - 10a$$

9. Combine like terms and write the resulting polynomial in descending order of degree.

$$4x - 2 + 5x + 6x^2 + 2$$

Since $4x$ and $5x$ are like terms, they combine to give $(4+5)x = 9x$. Similarly, combining the like terms -2 and 2 gives $-2 + 2 = 0$. The resulting polynomial is $6x^2 + 9x + 0$, which can be written just as $6x^2 + 9x$.

9. Combine like terms and write the resulting polynomial in descending order of degree.

$$c^2 + c^4 - 2 + 2c^2 - 3c^4$$

ADDITIONAL EXERCISES
Objective 1 Identify monomials.
For extra help, see Example 1 on page 153 of your text and the Section 3.2 lecture video.
Determine whether the expression is a monomial. Explain.

1. $\dfrac{4x}{y^3}$

2. 21

Objective 2 Identify the coefficient and degree of a monomial.
For extra help, see Example 2 on page 154 of your text and the Section 3.2 lecture video.
Identify the coefficient and degree of each monomial.

3. $7x^4 y^2$

4. -18

Objective 3 Identify polynomials and their terms.
For extra help, see Examples 3–4 on pages 155–156 of your text and the Section 3.2 lecture video.
Identify the terms in each polynomial and their coefficients.

5. $4y^5 + 13y^4 - y^3 - 12$

6. $16x^4 - 4x^2 + 2x - 1$

Indicate whether the expression is a monomial, binomial, trinomial, or has no special name.

7. $y^2 + 5y - 8$

8. $a^{15} - 2a^3 + 2a + 4$

Objective 4 Identify the degree of a polynomial.
For extra help, see Example 5 on page 157 of your text and the Section 3.2 lecture video.
Identify the degree of each polynomial.

9. $18d - 5 + 21d^3 + d^2$

10. $-2x^2 - 4x^3 + x^8$

Objective 5 Write polynomials in descending order of degree.
For extra help, see Example 6 on page 157 of your text and the Section 3.2 lecture video.
Write each polynomial in descending order of degree.

11. $-z^6 + 3z^2 - z + 15 - 4z^5$

12. $5a^2 - 9a^{11} + 4a^3 - 1$

Objective 6 Simplify polynomials in one variable by combining like terms.
For extra help, see Examples 7–8 on pages 158–159 of your text and the Section 3.2 lecture video.
Determine whether the monomials are like terms.

13. $4x$ and $2x^2$

14. $17b$ and $-b$

Combine like terms.

15. $6y^4 - 10y^4$

16. $21a^2 - 9a^2$

17. $7p + 3p$

18. $18a - 5a$

19. $x + 12x$

20. $-21y - 3y$

Combine like terms and write the resulting polynomial in descending order of degree.

21. $4x - 10 + 3x^5 - 2x - 2$

22. $15y^3 + 9y + 2y^2 - 9y + 3 - 21y^3$

23. $5 - 6v + 8v^2 + 20 + 7v - 8v^4$

24. $-7u^4 + 21u - 15 + 8u^4 - 2u^2 - 3u^3 - 13 + 2u$

25. $-2h - 5h^2 - 2h + 3 - 17h^2 + 4h^3 - 3 + 7h^3$

26. $t^6 + 8t^2 - 15 + 3t^4 - 6t^2 - 9t^3 - 2 + 3t + 5t^3$

Chapter 3 EXPRESSIONS AND POLYNOMIALS

3.3 Adding and Subtracting Polynomials

Learning Objectives
1   Add polynomials in one variable.
2   Subtract polynomials in one variable.

GUIDED EXAMPLES AND PRACTICE

Objective 1 Add polynomials in one variable.

Review this example for Objective 1:

1. Add and write the resulting polynomial in descending order of degree.

    $(6x+3)+(7x-9)$

    Combining like terms $6x$ and $7x$ gives $13x$ and combining like terms 3 and -9 gives -6. The resulting polynomial is $(6x+3)+(7x-9)=13x-6$.

Practice this exercise:

1. Add and write the resulting polynomial in descending order of degree.

    $(4k^2+2k-9)+(8k^2-3k+1)$

Objective 2 Subtract polynomials in one variable.

Review this example for Objective 2:

2. Subtract and write the resulting polynomial in descending order of degree.

    $(10n^2-5n+12)-(-4n^2+2n-1)$

    Change the operation from $-$ to $+$ and switch the sign in each term of the subtrahend.

    $(10n^2-5n+12)+(4n^2-2n+1)$.

    Combine like terms.

    $14n^2-7n+13$

Practice this exercise:

2. Subtract and write the resulting polynomial in descending order of degree.

    $(18y^2+11y+6)-(y^2+2y-1)$

ADDITIONAL EXERCISES

Objective 1 Add polynomials in one variable.

For extra help, see Examples 1–2 on pages 163–164 of your text and the Section 3.3 lecture video.

Add and write the resulting polynomial in descending order of degree.

1. $(8y - 22) + (3y + 11)$

2. $(7p^2 + 2p - 8) + (6p^2 - p - 20)$

3. $(8x^3 + 3x^2 - x + 18) + (5x^3 - 3x^2 + 7x - 1)$

4. $(2a^3 - 9a^2 + 5a + 1) + (4a^3 - 11a^2 + 3a + 7)$

5. $(12u^5 + 8u^3 - u^2 - 15) + (2u^5 - 6u^3 + 5u^2 + 12)$

6. $(9h^6 + 8h^4 - 3h - 17) + (-3h^6 - 15h^4 - 3h + 5)$

Write an expression in simplest form for the perimeter of each shape.

7.

$x + 13$

$x + 4$

8.

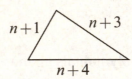

$n + 1$ $n + 3$

$n + 4$

Objective 2 Subtract polynomials in one variable.

For extra help, see Example 3 on page 165 of your text and the Section 3.3 lecture video.

Subtract and write the resulting polynomial in descending order of degree.

9. $(7x + 6) - (6x + 7)$

10. $(8p - 17) - (-2p + 12)$

11. $(21u^3 - 17u^2 + 9u + 2) - (-14u^3 + 12u^2 - 8u - 3)$

12. $(-3z^3 + 4z^2 - 9z + 8) - (2z^3 + 12z^2 - 15z - 7)$

13. $(6u^4 + 9u^2 - 8u - 12) - (8u^4 - 6u^2 + 7u + 5)$

14. $(19r^5 + 3r^3 - 2r - 1) - (-8r^5 - 2r^3 + 8r - 6)$

Chapter 3 EXPRESSIONS AND POLYNOMIALS

3.4 Exponent Rules; Multiplying Polynomials

Learning Objectives
1   Multiply monomials.
2   Simplify monomials raised to a power.
3   Multiply a polynomial by a monomial.
4   Multiply polynomials.

Key Terms

Use the terms listed below to complete each statement in Exercises 1–4.

    add   **multiply**   **subtract**   **divide**   **exponents**   **sign**   **base**

1.   To simplify a monomial raised to a power, evaluate the coefficient raised to that power and _____ each variable's exponent by the power.

2.   To multiply exponential forms that have the same base, add the _____ and keep the same _____.

3.   To multiply monomials, _____ the coefficients and _____ the exponents of the like variables.

4.   Conjugates are binomials that differ only in the _____ separating the terms.

GUIDED EXAMPLES AND PRACTICE

Objective 1 Multiply monomials.

Review this example for Objective 1:

1. Multiply.

$$-t^3(2t)(-5t^4)$$

Multiply coefficients and add exponents of the like bases, t.

$$=(-1)\cdot 2\cdot(-5)t^{3+1+4}$$

$$=10t^8 \qquad \text{Simplify}$$

Practice this exercise:

1. Multiply.

$$x^{12}\cdot x^5$$

Objective 2 Simplify monomials raised to a power.

Review this example for Objective 2:

2. Simplify.

$$\left(3h^9\right)^2$$

Evaluate the coefficient, 3, raised to the power, 2, and multiply the variable's exponent, 9, by the power, 2.

$$\left(3h^9\right)^2 = \left(3\right)^2 h^{9 \cdot 2} = 9h^{18}$$

Practice this exercise:

2. Simplify.

$$\left(4k^7\right)^6$$

Objective 3 Multiply a polynomial by a monomial.

Review this example for Objective 3:

3. Multiply.

$$3p\left(2p - 5\right)$$

Evaluate the coefficient, 3, raised to the power, 2, and multiply the variable's exponent, 9, by the power, 2.

$$\left(3h^9\right)^2 = \left(3\right)^2 h^{9 \cdot 2} = 9h^{18}$$

Practice this exercise:

3. Multiply.

$$-7\left(4x - 5\right)$$

Objective 4 Multiply polynomials.

Review these examples for Objective 4:

4. Multiply.

$$\left(x + 6\right)\left(x - 1\right)$$

Distribute x to x and -1 and distribute 6 to x and -1.

$$\left(x + 6\right)\left(x - 1\right)$$

$$= x \cdot x + x \cdot \left(-1\right) + 6 \cdot x + 6 \cdot \left(-1\right)$$

$$= x^2 - x + 6x - 6 \qquad \text{Multiply.}$$

$$= x^2 + 5x - 6 \qquad \text{Combine like terms.}$$

Practice these exercises:

4. Multiply.

$$\left(2b - 3\right)\left(2b + 3\right)$$

5. Multiply.

$$(x+2)(x^2-3x+4)$$

Distribute both x and 2 to each term in (x^2-3x+4).

$$(x+2)(x^2-3x+4)$$

$$= x\cdot x^2 - x\cdot(-3x) + x\cdot 4 + 2\cdot x^2 + 2\cdot(-3x) + 2\cdot 4$$

$$= x^3 + 3x^2 + 4x + 2x^2 - 6x + 8 \qquad \text{Multiply.}$$

$$= x^3 + 5x^2 - 2x + 8 \qquad \text{Combine like terms}$$

5. Multiply.

$$(y+5)(2y^2-6y+3)$$

ADDITIONAL EXERCISES

Objective 1   Multiply monomials.

For extra help, see Examples 1–2 on pages 169–170 of your text and the Section 3.4 lecture video.

Multiply.

1. $p^2 \cdot p$

2. $x^3(-2x^5)(-6x^4)$

3. $3y^2 \cdot 8y^3$

4. $7w^4(-2w)$

Objective 2   Simplify monomials raised to a power.

For extra help, see Example 3 on page 172 of your text and the Section 3.4 lecture video.

Simplify.

5. $\left(-3w^5\right)^3$

6. $3x^2 \cdot (5x^3)^2$

Objective 3   Multiply a polynomial by a monomial.

For extra help, see Example 4 on page 173 of your text and the Section 3.4 lecture video.

Multiply.

7. $6(3y-1)$

8. $5a^2\left(-2a^2-a\right)$

9. $-4n^5\left(2n^2-3n+1\right)$

10. $-8w^4\left(2w^2+5w-4\right)$

Objective 4 Multiply polynomials.
For extra help, see Examples 5–11 on pages 174–178 of your text and the Section 3.4 lecture video.
Multiply.

11. $(u-3)(u+4)$

12. $(2c+7)(4c-1)$

13. $(2x-3y)(5x-y)$

14. $(8m-5n)(7m+3n)$

15. $(3s+11t)(2s-7t)$

16. $(9t-2)(3t-5)$

17. $(a-7)(a+7)$

18. $t^2(t+3)(3t-1)$

19. $-3b^2(2b+4)(3b-2)$

20. $2x^3(2x+3)(3x-2)$

Write an expression.

21. *Write an expression for the area of the rectangle.*

$3y$

$7y+4$

22. *Write an expression for the volume of the box.*

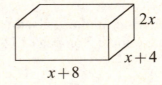

$2x$

$x+4$

$x+8$

Chapter 3 EXPRESSIONS AND POLYNOMIALS

3.5 Prime Numbers and GCF

Learning Objectives
1   Determine whether a number is prime, composite, or neither.
2   Find the prime factorization of a given number.
3   Find all factors of a given number.
4   Find the greatest common factor of a given set of numbers by listing.
5   Find the greatest common factor of a given set of numbers using prime factorization.
6   Find the greatest common factor of a set of monomials.

Key Terms

Use the vocabulary terms listed below to complete each statement in Exercises 1–4.

natural	**prime**	**composite**	**factor**
greatest	**quotient**	**least**	

1.   A prime factorization is written with only _____ factors.

2.   A _____ number has exactly one other factor than itself.

3.   The GCF is the _____ number that divides all given numbers with no remainder.

4.   A _____ number has factors other than one and itself.

GUIDED EXAMPLES AND PRACTICE

Objective 1 Determine whether a number is prime, composite, or neither.

Review these examples for Objective 1:

1.   Determine if the number is prime, composite, or neither.

    6

    Divide by the list of primes.
    Is 6 divisible by 2? Yes, since $2 \cdot 3 = 6$. Since 2 is a factor of 6 which is neither 1 nor 6, 6 is not prime.

Practice these exercises:

1.   Determine if the number is prime, composite, or neither.

    39

2. Determine if the number is prime, composite, or neither.

17

Divide by the list of primes.
Is 17 divisible by 2? No, since 17 is odd.
Is 17 divisible by 3?

$$\begin{array}{r} 5 \\ 3)\overline{17} \\ -15 \\ \hline 2 \end{array}$$

No, 17 is not divisible by 3.
Because the quotient, 5 is
greater than the divisor, 3,
we must go on to the next prime

Is 17 divisible by 5?

$$\begin{array}{r} 3 \\ 5)\overline{17} \\ -15 \\ \hline 2 \end{array}$$

No, 17 is not divisible by 5.
Because the quotient, 3 is
less than the divisor, 5,
we can stop and conclude that
17 is a prime number.

2. Determine if the number is prime, composite, or neither.

23

Objective 2 Find the prime factorization of a given number.

Review this example for Objective 2:

3. Find the prime factorization. Write the answer in exponential form.

405

Use a factor tree.

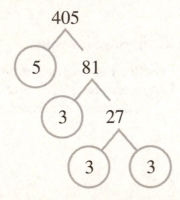

The prime factorization is $405 = 3^4 \cdot 5$.

Practice this exercise:

3. Find the prime factorization. Write the answer in exponential form.

200

Objective 3 Find all factors of a given number.

Review this example for Objective 3:

4. List all factors of each number.
    36

    Divide by the natural numbers $1, 2, 3, \ldots$ until
    we have all of the factors
    $1 \cdot 36$
    $2 \cdot 18$
    $3 \cdot 12$
    $4 \cdot 9$
    $6 \cdot 6$
    Because 36 is not divisible by 5, we have
    found all factors. The list of factors is
    1, 2, 3, 4, 6, 9, 12, 18, 36.

Practice this exercise:

4. List all factors of each number.
    54

Objective 4 Find the greatest common factor of a given set of numbers by listing.

Review this example for Objective 4:

5. Find the GCF by listing.
    189 and 36

    List the factors of each number.
    Factors of 189: 1, 3, 7, 9, 21, 27, 63, 189
    Factors of 36: 1, 3, 4, 6, 9, 12, 18, 36
    Since 9 is the greatest number in both of these
    lists, GCF = 9.

Practice this exercise:

5. Find the GCF by listing.
    27 and 26

Objective 5 Find the greatest common factor of a given set of numbers using prime factorization.

Review this example for Objective 5:

6. Find the GCF using prime factorization.
    135 and 225

    First, find the prime factorizations of 135
    and 225.
    $135 = 3^3 \cdot 5$
    $225 = 3^2 \cdot 5^2$
    The common prime factors are 3 and 5. For 3,
    the least exponent is 2. For 5, the least
    exponent is 1. The GCF is the product
    $GCF = 3^2 \cdot 5 = 9 \cdot 5 = 45$.

Practice this exercise:

6. Find the GCF using prime
    factorization.
    343 and 8

Objective 6 Find the greatest common factor of a set of monomials.

Review this example for Objective 6:

7. Find the GCF of the monomials.

$20x$ and $45x^2$

Find the prime factorization of each monomial. Treat the variable, x, as a prime factor.

$20x = 2^2 \cdot 5 \cdot x$

$45x^2 = 3^2 \cdot 5 \cdot x^2$

The common factors are 5 and x. For 5, both exponents are 1. For x, the least exponent is 1. The GCF is the product GCF $= 5 \cdot x = 5x$.

Practice this exercise:

7. Find the GCF of the monomials.

$110x^5$, $70x^6$, and $60x^8$

ADDITIONAL EXERCISES

Objective 1 Determine whether a number is prime, composite, or neither.

For extra help, see Example 1 on page 184 of your text and the Section 3.5 lecture video.

Determine if the number is prime, composite, or neither.

1. 37

2. 1

3. 125

4. 243

Objective 2 Find the prime factorization of a given number.

For extra help, see Example 2 on page 185 of your text and the Section 3.5 lecture video.

Find the prime factorization. Write the answer in exponential form.

5. 512

6. 675

7. 2205

8. 138

Objective 3 Find all factors of a given number.

For extra help, see Example 3 on page 186 of your text and the Section 3.5 lecture video.

List all factors of each number.

9. 375

10. 225

Objective 4 Find the greatest common factor of a given set of numbers by listing.
For extra help, see Example 4 on page 187 of your text and the Section 3.5 lecture video.
Find the GCF by listing.

11. 22 and 29

12. 24 and 225

Objective 5 Find the greatest common factor of a given set of numbers using prime factorization.
For extra help, see Examples 5–6 on pages 189–190 of your text and the Section 3.5 lecture video.
Find the GCF using prime factorization.

13. 40, 100, and 20

14. 20, 28, and 24

Objective 6 Find the greatest common factor of a set of monomials.
For extra help, see Example 7 on page 190 of your text and the Section 3.5 lecture video.
Find the GCF of the monomials.

15. $60x^3$, $54x^6$, and $294x^6$

16. $17x^3$, $24x$, and $6x^7$

Chapter 3 EXPRESSIONS AND POLYNOMIALS

3.6 Exponent Rules; Introduction to Factoring

Learning Objectives
1   Divide monomials.
2   Divide a polynomial by a monomial.
3   Find an unknown factor.
4   Factor the GCF out of a polynomial.

Key Terms
Use the vocabulary terms listed below to complete each statement in Exercises 1–3.

    divisor's   **dividend's**   **exponent**   **product**   **quotient**

1.   A number or expression is said to be in factored form if it is written as the _____ of factors.

2.   To divide exponential forms that have the same base, subtract the _____ exponent from the _____ exponent and keep the same base.

3.   To factor a monomial GCF out of a polynomial, write the polynomial as a product of the GCF and the _____ of the polynomial and the GCF.

GUIDED EXAMPLES AND PRACTICE

Objective 1   Divide monomials.

Review this example for Objective 1:	**Practice this exercise:**
1. Divide. $x^7 \div x^3$ Subtract the divisor's exponent from the dividend's exponent and keep the same base. $x^7 \div x^3 = x^{7-3} = x^4$	1. Divide. $15x^6 \div 5x^4$

Objective 2   Divide a polynomial by a monomial.

Review this example for Objective 2:	**Practice this exercise:**
2. Divide. $\dfrac{-14x^5 + 35x}{7x}$	2. Divide. $\dfrac{18x^8 - 24x^6 + 15x}{3x}$

Divide each term in $-14x^5 + 35x$ by the monomial $7x$.

$$\frac{-14x^5 + 35x}{7x}$$

$$= \frac{-14x^5}{7x} + \frac{35x}{7x}$$

Divide the coefficients, subtract the exponents of the like bases, and simplify.

$$= \frac{-14}{7}x^{5-1} + \frac{35}{7}x^{1-1}$$

$$= -2x^4 + 5 \qquad \text{Simplify.}$$

Objective 3 Find an unknown factor.

Review these examples for Objective 3:

3. Find the unknown factor.

$$78x^4 = 6x^2 \cdot (?)$$

Solve for the missing factor by writing a related division sentence.

$$(?) = \frac{78x^4}{6x^2}$$

Divide the coefficients, subtract the exponents of the like bases, and simplify.

$$(?) = \frac{78}{6}x^{4-2} = 13x^2$$

4. Find the unknown factor.

$$-35a^6 + 14a^5 = 7a^5 \cdot (?)$$

Solve for the missing factor by writing a related division sentence.

$$(?) = \frac{-42a^6 + 18a^5}{6a^5}$$

Divide each term in $-42a^6 + 18a^5$ by the monomial $6a^5$.

$$= \frac{-42a^6}{6a^5} + \frac{18a^5}{6a^5}$$

Divide the coefficients, subtract the exponents of the like bases, and simplify.

$$= \frac{-42}{6}a^{6-5} + \frac{18}{6}a^{5-5} = -7a + 3$$

Practice these exercises:

3. Find the unknown factor.

$$-42a^8 = (?) \cdot (-6a^7)$$

4. Find the unknown factor.

$$125x^3 + 15x^4 = (?) \cdot (5x^2)$$

Objective 4 Factor the GCF out of a polynomial.

Review this example for Objective 4:

5. Factor.

$4x^2 + 36x$

First, find the GCF of $4x^2$ and $36x$, which is GCF $= 4x$. Rewrite the polynomial as the product of the GCF and the quotient of the polynomial and GCF.

$$= 4x\left(\frac{4x^2 + 36x}{4x}\right)$$

Divide each term in the polynomial by the GCF and simplify.

$$= 4x\left(\frac{4x^2}{4x} + \frac{36x}{4x}\right)$$

$$= 4x(x + 9)$$

Practice this exercise:

5. Factor.

$2c + 32$

ADDITIONAL EXERCISES

Objective 1 Divide monomials.

For extra help, see Examples 1–3 on pages 194–196 of your text and the Section 3.6 lecture video.

Divide.

1. $-63x^4 \div 7x^2$

2. $9x^4 \div 3x^4$

3. $\dfrac{-28y^7}{-7y^3}$

4. $\dfrac{110r^{10}}{-5r^7}$

Objective 2 Divide a polynomial by a monomial.

For extra help, see Example 4 on page 197 of your text and the Section 3.6 lecture video.

Divide.

5. $(30x + 15) \div 5$

6. $(16x^6 - 10x^3) \div (2x)$

7. $\dfrac{63x^9 - 35x^4 + 35x^2}{-7x^2}$

8. $\dfrac{54y^7 + 42y^5 - 6y^2}{3y^2}$

Objective 3 Find an unknown factor.
For extra help, see Examples 5–6 on page 198 of your text and the Section 3.6 lecture video.
Find the unknown factor.

9. $40y^7 = 5y^6 \cdot (?)$

10. $-96x^3 = 6x^3 \cdot (?)$

11. $-24d^7 + 30d^3 = 6d^2 \cdot (?)$

12. $5x^4 - 45x^3 + 20x^2 = 5x^2 \cdot (?)$

Find the missing side length.

13. $A = 45y^2$

14. $V = 24m^8$

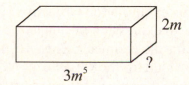

Objective 4 Factor the GCF out of a polynomial.
For extra help, see Example 7 on page 200 of your text and the Section 3.6 lecture video.
Factor.

15. $5a - 15$

16. $9x^2 + 45x$

17. $2b - 7b^4$

18. $5x^7 - 10x^5$

19. $16x^3 - 12x^2 + 4x$

20. $18x^3 - 12x^2 + 6x$

21. $7x^6 - 19x^4 + 5x^3$

22. $18x^5 - 5x^3 + 11x^2$

Chapter 3 EXPRESSIONS AND POLYNOMIALS

3.7 Additional Applications and Problem Solving

Learning Objectives
1 Solve polynomial problems involving perimeter, area, and volume.
2 Solve surface area problems.
3 Solve problems involving a falling object.
4 Solve net-profit problems.

GUIDED EXAMPLES AND PRACTICE

Objective 1 Solve polynomial problems involving perimeter, area, and volume.

Review this example for Objective 1:

1. Write an expression in simplest form for the perimeter. Then use the given value for the variable to calculate the value of the perimeter.

$4x - 1$

$2x + 2$

$x = 7$ centimeters

Because perimeter is the sum of the side lengths, we add the lengths of the sides together.

Perimeter $= 2(4x - 1) + 2(2x + 2)$

$= 8x - 2 + 4x + 4$

$= 12x + 2$

Now evaluate $12x + 2$ when $x = 7$.

$12(7) + 2 = 84 + 2 = 86$ centimeters

Practice this exercise:

1. Write an expression in simplest form for the perimeter. Then use the given value for the variable to calculate the value of the perimeter.

$7x - 9$

$5x + 10$

$x = 15$ feet

Objective 2 Solve surface area problems.

Review this example for Objective 2:

2.   A gift box 18 centimeters long by 9 centimeters wide by 2 centimeters high is to be wrapped in paper. What is the total area of paper that will be needed if there is no overlapping?

The surface area of a box is
$SA = 2(lw + lh + wh)$ where l is length, w is width, and h is height. In this equation replace l with 18, w with 9, and h with 2.

$SA = 2(18 \cdot 9 + 18 \cdot 2 + 9 \cdot 2)$

$\quad = 2(162 + 36 + 18)$

$\quad = 2(216)$

$\quad = 432$

The total area of paper needed is 432 cm^2.

Practice this exercise:

2.   A large cube of ice is 3 feet along each side. What is the total surface area of the cube?

Objective 3 Solve problems involving a falling object.

Review this example for Objective 3:

3.   The height of a building is 1388 feet. If an object was dropped from the top of the building, what would be the height of the object after 5 seconds?

The height of the object after t seconds is $h = -16t^2 + h_0$ where h_0 is the initial height of the object. Since the object is dropped from the top of the building $h_0 = 1388$.

Plugging in 5 for t, we see that the height of the object after 5 seconds is $h = -16(5)^2 + 1388 = 988$ feet.

Practice this exercise:

3.   The surface of a road going across a bridge is 1152 feet above the river below it. If a rock rolls off the road on the bridge, how far above the water is the rock after 6 seconds?

Objective 4 Solve net-profit problems.

Review this example for Objective 4:

4. A business sells clocks and lamps. If x represents the number of clocks and y represents the number of lamps, then $25x + 6y$ describes the revenue from the sale of the two products. The polynomial $5x + y + 1$ describes the cost of producing the two products.

   a. Write an expression in simplest form for net profit.

   b. If the business sold 3396 clocks and 2654 lamps in a month, what was the net profit for the month?

   a. Net profit is revenue, R, minus cost, C. Since $R = 25x + 6y$ and $C = 5x + y + 1$, the net profit is

$$N = R - C = (25x + 6y) - (5x + y + 1)$$
$$= 20x + 5y - 1$$

   b. $N = 20(3396) + 5(2754) - 1$
$$= 81689$$

Practice this exercise:

4. A footwear company sells shoes and sandals. If x represents the number of shoes and y represents the number of sandals, then $36x + 8y$ describes the revenue from the sale of shoes and sandals (in dollars). The polynomial $3x + 2y + 10$ describes the cost of producing the two products (in dollars).

   a. Write an expression in simplest form for net profit.

   b. If the business sold 432 shoes and 220 sandals, what was the net profit?

ADDITIONAL EXERCISES

Objective 1 Solve polynomial problems involving perimeter, area, and volume.

For extra help, see Example 1 on page 203 of your text and the Section 3.7 lecture video.

Write an expression in simplest form for the perimeter. Then use the given value for the variable to calculate the value of the perimeter.

1.

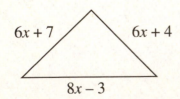

$x = 3$ yards

2.

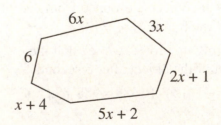

$x = 3$ meters

Write an expression in simplest form for the area. Then use the given value for the variable to calculate the value of the area.

3.

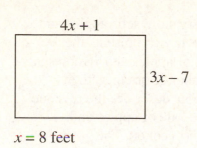

4x + 1

3x − 7

x = 8 feet

4.

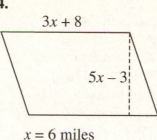

3x + 8

5x − 3

x = 6 miles

Write an expression in simplest form for the volume. Then use the given value for the variable to calculate the value of the volume.

5.

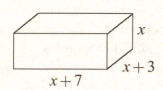

x

x + 3

x + 7

x = 2 inches

6.

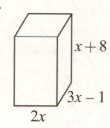

x + 8

3x − 1

2x

x = 4 centimeters

Objective 2 Solve surface area problems.
For extra help, see Example 2 on page 205 of your text and the Section 3.7 lecture video.
Solve.

7.   A brick 12 centimeters long by 6 centimeters wide by 5 centimeters high is to be painted on every side. What is the total area to be painted?

8.   A cereal box is printed on all sides with color and type. If the box is 8 inches long by 2 inches wide by 10 inches high, what is the total area that is printed?

Objective 3 Solve problems involving a falling object.
For extra help, see Example 3 on page 205 of your text and the Section 3.7 lecture video.
Solve.

9.   A skydiver jumps from a plane at 15,000 feet and deploys his parachute after 25 seconds of free fall. What is his altitude at the time he deploys his parachute.

10. The height of a building is 560 meters. If an object was dropped from the top of the building, what would be the height of the object after 4 seconds?

Objective 4 Solve net-profit problems.
For extra help, see Example 4 on page 206 of your text and the Section 3.7 lecture video.
Solve.

11. Clarice sews blankets in three sizes; small, medium, and large. She sells the small blankets for $10, medium for $12, and large for $16. Each small blankets costs $4 to make; each medium, $6; and each large, $7.

    a. Write polynomial that describes the revenue she receives from selling all three blankets.

    b. Write polynomial that describes the cost to produce all three blankets.

    c. Write an expression in simplest for her net.

    d. In one day, she sells 14 small blankets, 6 medium blankets, and 12 large blankets. Find her net profit for the day.

12. A grocery store sells apples and oranges. If a represents the number of apples and b represents the number of oranges, then $32a + 55b$ describes the revenue from the sale of apples and oranges (in cents). The polynomial $25a + 30b + 5000$ describes the cost of buying the two products from fruit farmers (in cents).

    a. Write an expression in simplest form for net profit.

    b. If the store sold 1255 apples and 588 oranges in a month, what was the net profit for that month in dollars?

Name: Date:
Instructor: Section:

Chapter 4 EQUATIONS

4.1 Equations and Their Solutions

Learning Objectives
1   Differentiate between an expression and an equation.
2   Check to see whether a number is a solution for an equation.

Key Terms
Use the vocabulary terms listed below to complete each statement in Exercises 1–4.

equation   expression   equate   solve   solution

1. To check a solution, simplify each side of the _____ separately to determine if it is true or false.

2. The goal with an equation is to _____ it.

3. An equation has a(n) _____ on each side of the equal sign.

4. To check a(n) _____ of an equation, replace the variable(s) with the number and simplify both sides of the equation.

GUIDED EXAMPLES AND PRACTICE

Objective 1 Differentiate between an expression and an equation.

Review these examples for Objective 1:
1. Determine whether the following is an expression or an equation.
$$7(x+5)-6(2x-5)+13$$

It is an expression because it does not contain an equal sign.

2. Determine whether the following is an expression or an equation.
$$9(x+3)-5(9x-5)=17$$

It is an equation because it does contain an equal sign.

Practice these exercises:
1. Determine whether the following is an expression or an equation.
$$5x^2+17=2x$$

2. Determine whether the following is an expression or an equation.
$$2x^2+3xy-7y$$

Objective 2 Check to see whether a number is a solution for an equation.

Review these examples for Objective 2:	Practice these exercises:

Review these examples for Objective 2:

3. Check to see if the given number is a solution for the given equation.
    $x+17=40$; check $x=16$

    $x+17=40$

    $16+17\overset{?}{=}40$ Replace.

    $33=40$ Simplify.

    False; so 16 is not a solution.

4. Check to see if the given number is a solution for the given equation.
    $6x^2-60=36$; check $x=4$

    $6x^2-60\overset{?}{=}36$

    $6\left(4^2\right)-60\overset{?}{=}36$ Replace.

    $6(16)-60\overset{?}{=}36$ Simplify.

    $36=36$ True; so 4 is

    a solution.

Practice these exercises:

3. Check to see if the given number is a solution for the given equation.
    $x+14=35$; check $x=21$

4. Check to see if the given number is a solution for the given equation.
    $32x^2-76=54$; check $x=2$

ADDITIONAL EXERCISES

Objective 1 Differentiate between an expression and an equation.

For extra help, see Examples 2–3 on pages 225–226 of your text and the Section 4.1 lecture video.

Indicate whether each of the following is an expression or an equation.

1. $8x^2-4x+1=0$

2. $9(x-2)+7(x+5)-3$

3. $9+2(x-5)+x^2$

4. $6y^2-3(5+y)+11=42$

Objective 2 Check to see whether a number is a solution for an equation.

For extra help, see Example 2 on page 226 of your text and the Section 4.1 lecture video.

Check to see if the given number is a solution for the given equation.

5. $6x=-15$
    check $x=-6$

6. $7x+5=40$
    check $x=5$

7. $7x + 7 = 56$

check $x = 7$

8. $5x + 7 = 92$

check $x = 17$

9. $5x + 5 = 56$

check $x = 10$

10. $9x + 7 = 56$

check $x = 5$

11. $9x + 7 = 10x$

check $x = 9$

12. $2x^2 - 75 = 53$

check $x = 8$

13. $7x + 7 = 8x$

check $x = 8$

14. $6x^2 - 135 = 15$

check $x = 5$

15. $9(y - 4) = -2y + 52$

check $y = 8$

16. $7(x - 3) = -2x + 60$

check $x = 9$

17. $9(z - 4) = -3z + 52$

check $z = 9$

18. $2(x - 4) = -6x + 56$

check $x = 5$

Chapter 4 EQUATIONS

4.2 The Addition Principle of Equality

Learning Objectives
1   Determine whether a given equation is linear.
2   Solve linear equations in one variable using the addition principle of equality.
3   Solve application problems.

Key Terms
Use the vocabulary terms listed below to complete each statement in Exercises 1–4.

linear	degree	monomial	distributive	solution
nonlinear	multiply	inequality	subtract	add

1. Adding the same amount to both sides of an equation does not change the equation's

    _____.

2. If an equation in one variable has degree greater than 1, it is a(n) _____
    equation.

3. To clear a term that is added to one side of an equation, _____ the
    additive inverse of the term to both sides of the equation.

4. If an equation has parentheses, use the _____ property to clear the
    parentheses.

GUIDED EXAMPLES AND PRACTICE

Objective 1 Determine whether a given equation is linear.

Review this example for Objective 1:
1. Determine whether the given equation
    is linear.
    $$4x^2 + 2(x-3) = 11$$

    $4x^2 + 2(x-3) = 11$ is nonlinear

    because it has a variable term, $4x^2$,
    with a degree other than 1. The degree
    of $4x^2$ is 2.

Practice this exercise:
1. $3x + 5(x-2) = 7$

         Copyright © 2013 Pearson Education, Inc.

Objective 2   Solve linear equations in one variable using the addition principle of equality.

Review these examples for Objective 2:

2.   Solve and check

$x + 13 = 15$

$x + 13 = 15$

$\underline{-13 \quad -13}$ Subtract 13 from both sides

$x + 0 = 2$ The sum of x and 0 is x.

$x = 2$

Check:

$x + 13 = 15$

$2 + 13 = 15$ Replace x with 2.

$15 = 15$ True; so 2 is the

solution.

3.   Solve and check

$11 + 7 = 4(y + 3) - 3y$

$11 + 7 = 4(y + 3) - 3y$

$11 + 7 = 4y + 12 - 3y$ Distribute to clear parentheses.

$18 = y + 12$ Simplify each side.

$\underline{-12 \qquad -12}$ Subtract 12.

$6 = y$

Check:

$11 + 7 = 4(y + 3) - 3y$

$11 + 7 = 4((6) + 3) - 3(6)$ Replace y with 6 in the original equation.

$11 + 7 = 4(9) - 3(6)$ Simplify each side.

$11 + 7 = 36 - 18$

$18 = 18$ True; so 6 is the

solution.

Practice these exercises:

2.   Solve and check

$y + 12 = -8$

3.   Solve and check

$2(h - 3) - h = 21 - 18$

Objective 3 Solve application problems.

Review this example for Objective 3:

4.   Translate to an equation, then solve. Suppose Aida is planning to buy a new car. The down payment is $4444. She has $238 saved toward that amount. How much more does she need?

Adding the amount needed to her current amount, $238, should equal the total needed. Let x represent the unknown addend.

Current amount + needed amount = 4444

$$238 \quad + \quad x \quad = 4444$$

Solve:

$$238 + x = 4444$$

$$\underline{-238 \qquad -238} \quad \text{Subtract 238 from}$$
$$\qquad\qquad\qquad\qquad \text{both sides.}$$

$$0 + x = 4206$$

$$x = 4206$$

Practice this exercise:

4.   A golfer's average annual expense for greens fees, cart rentals, and golf balls is $346. If he spends $168 for greens fees and $148 for cart rentals, how much does he spend for golf balls?

ADDITIONAL EXERCISES

Objective 1 Determine whether a given equation is linear.

For extra help, see Example 1 on page 229 of your text and the Section 4.2 lecture video.

Determine whether the given equation is linear.

1. $5x + 2(x-1) = 9 - x$ 2. $n^3 + 9 = y - 3$

Objective 2 Solve linear equations in one variable using the addition principle of equality.

For extra help, see Examples 2–4 on pages 231–232 of your text and the Section 4.2 lecture video.

Solve and check.

3. $-9 = n - 21$ 4. $15 = q - 19$

5. $4t + 12 - 3t = 18$

6. $8m + 12 - 7m = 14 - 3$

7. $17 + 8 = 3w - 5 - 2w$

8. $10x - 3(3x + 2) = -15 + 14$

Objective 3 Solve application problems.
For extra help, see Example 5 on page 232 of your text and the Section 4.2 lecture video.
Translate to an equation, then solve.

9. The perimeter of the triangle is 86 millimeters. Find the length of the missing side.

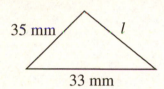

35 mm   *l*

33 mm

10. Marc is tiling his dining room with tiles that measure 1 foot by 1 foot. The room is 180 square feet. He has already purchased 125 tiles. How many more tiles does he need?

Chapter 4 EQUATIONS

4.3 The Multiplication Principle of Equality

Learning Objectives
1   Solve equations using the multiplication principle of equality.
2   Solve equations using both the addition and multiplication principles of equality.
3   Solve application problems.

GUIDED EXAMPLES AND PRACTICE

Objective 1 Solve equations using the multiplication principle of equality.

Review this example for Objective 1:

1. Solve using the multiplication principle, then check.
    $$7y = -21$$

    $$\frac{7y}{7} = \frac{-21}{7} \quad \text{Divide both sides by 7.}$$

    $$y = -3$$

    Check:
    $$7y = -21$$

    $$7(-3) \overset{?}{=} -21 \quad \text{In the original equation,}$$
    $$\text{replace y with } -3.$$

    $$-21 = -21 \quad \text{True; so } -3 \text{ is the solution.}$$

Practice this exercise:

1. Solve using the multiplication principle, then check.
    $$8y = 16$$

Objective 2 Solve equations using both the addition and multiplication principles of equality.

Review these examples for Objective 2:

2. Solve using the multiplication and addition principles of equality, then check.
    $$4x + 8 = 20$$

Practice these exercises:

2. Solve using the multiplication and addition principles of equality, then check.
    $$5m - 7 = -12$$

$$4x + 8 = 20$$

$\underline{-8 \quad -8}$ Subtract 8 from both sides.

$$4x = 12$$

$\overline{4 \quad 4}$ Divide both sides by 4.

$$x = 3$$

Check:

$$4x + 8 = 20$$

$$4(3) + 8 \overset{?}{=} 20 \quad \text{Replace } x \text{ with 3.}$$

$$12 + 8 \overset{?}{=} 20 \quad \text{Simplify.}$$

$$20 = 20 \quad \text{True; so 3 is the solution.}$$

3. Solve using the multiplication and addition principles of equality, then check.
$$6x - 15 = 4x - 13$$

$$6x - 15 = 4x - 13$$

$\underline{-4x \qquad -4x}$ Subtract $4x$ from both sides.

$$2x - 15 = 0 - 13$$

$$2x - 15 = -13$$

$\underline{+15 \quad +15}$ Add 15 to both sides.

$$x2 + 0 = 2$$

$$x2 = 2$$

$\overline{2 \quad 2}$ Divide both sides by 2.

$$x = 1$$

Check:

$$6x - 15 = 4x - 13$$

$$6(1) - 15 \overset{?}{=} 4(1) - 13 \quad \text{Replace } x \text{ with 1.}$$

$$6 - 15 \overset{?}{=} 4 - 13 \quad \text{Simplify.}$$

$$-9 = -9 \quad \text{True; so 1 is the solution.}$$

3. Solve using the multiplication and addition principles of equality, then check.
$$8c + 15 = -19 - 9c$$

4. Solve using the multiplication and addition principles of equality, then check.

$$3(u-2)=5(u-1)+1$$

$$3(u-2)=5(u-1)+1$$

$\quad 3u-6=5u-4 \qquad$ Distribute 3 and 5.

$\quad \underline{-3u \qquad -3u} \qquad$ Subtract $3u$ from

both sides.

$\quad 0-6=2u-4$

$\quad -6=2u-4$

$\quad \underline{+4 \qquad +4} \qquad$ Add 4 to both sides.

$\quad -2=2u-0$

$\quad -2=2u$

$\quad \dfrac{}{2} \ \dfrac{}{2} \qquad$ Divide by 2.

$\quad -1=u$

Check:

$$3(u-2)=5(u-1)+1$$

$$3((-1)-2)\overset{?}{=}5((-1)-1)+1 \quad \text{Replace } u$$

$$\text{with } -1.$$

$$3(-3)\overset{?}{=}5(-2)+1 \qquad \text{Simplify.}$$

$$-9\overset{?}{=}-10+1$$

$$-9=-9 \qquad\qquad \text{True; so } -1 \text{ is}$$

$$\text{the solution.}$$

4. Solve using the multiplication and addition principles of equality, then check.

$$-9(v+1)-3=-2v+11-3(v+1)$$

Objective 3 Solve application problems.

Review this example for Objective 3:

5. Solve for the unknown amount. The area of a rectangular room is 156 square feet. The width is 13 feet. What is the length?

$$A = lw$$

$156 = l \cdot 13$ Replace A with 156

              and w with 13.

$156 = 13l$

$\dfrac{13}{\ } \quad \dfrac{13}{\ }$ Divide by 13.

$12 = l$

The length is 12 feet.

Practice this exercise:

5. The Klein family is planning a 300 mile trip. If they travel at an average speed of 60 miles per hour, what will be their travel time?

ADDITIONAL EXERCISES

Objective 1 Solve equations using the multiplication principle of equality.

For extra help, see Example 1 on page 232 of your text and the Section 4.3 lecture video.

Solve using the multiplication principle, then check.

1. $-5a = 25$

2. $7g = -35$

3. $-13r = -39$

4. $-26y = 104$

Objective 2 Solve equations using both the addition and multiplication principles of equality.

For extra help, see Examples 2–5 on pages 238–241 of your text and the Section 4.3 lecture video.

Solve using the multiplication and addition principles of equality, then check.

5. $18 = 2s + 14$

6. $-24 = 3y + 21$

7. $17 = -3r - 7$

8. $81 = 30 - 17n$

9. $-6q + 13 - 3q = 17 - q - 20$

10. $13y - 4 - 14y = 16 + 3y - 12$

11.

$-3p - 19 + 7p + 12 = 17p + 16 - p + 1$

12. $14h + 4 - 9h - 1 = h + 7 - 2h + 2$

13.

$7(x + 4) - 2(x - 1) = -4(x - 3) + 36$

14.

$-3(m - 10) - 2(m + 4) = -2 + 6m - (m + 16)$

Objective 3 Solve application problems.
For extra help, see Examples 6–7 on page 242 of your text and the Section 4.3 lecture video.
Solve for the unknown amount.

15. The perimeter of a rectangle is 94 centimeters and the length is 27 centimeters. Find the width of the rectangle.

16. The voltage of a circuit is -144 volts. If the current is -6 amps, find the resistance in ohms. (Use $V = ir$.)

Chapter 4 EQUATIONS

4.4 Translating Word Sentences to Equations

Learning Objectives
1   Translate sentences to equations using key words; then solve.

GUIDED EXAMPLES AND PRACTICE

Objective 1 Translate sentences to equations using key words; then solve.

Review these examples for Objective 1:

1. Translate to an equation; then solve. Ten more than a number is equal to negative six.

   Translate:

   Ten more than a number is equal to negative six.

   $$10 \quad + \quad n \quad = \quad -6$$

   Solve:
   $$10 + n = -6$$
   $$\underline{-10 \qquad -10} \quad \text{Subtract 10 from}$$
   $$\text{both sides.}$$
   $$0 + n = -16$$
   $$n = -16$$

2. Translate to an equation; then solve. Six more than the product of five and x yields forty-six.

   Translate:

   Six more than the product of five and x yields forty-six.

   $$6 \quad + \qquad \qquad 5 \quad \cdot \quad x \quad = \quad 46$$

   Solve:
   $$6 + 5x = 46$$
   $$\underline{-6 \qquad \quad -6} \quad \text{Subtract 6 from}$$
   $$\text{both sides.}$$
   $$0 + 5x = 40$$
   $$5x = 40$$
   $$\frac{5}{5} \quad \frac{40}{5} \quad \text{Divide by 5.}$$
   $$x = 8$$

Practice these exercises:

1. Translate to an equation; then solve. The difference of a number and seven is equal to fourteen.

2. Translate to an equation; then solve. Fourteen subtracted from the product of negative two and y is sixteen.

3. Translate to an equation; then solve. Four times the difference of *b* and six is equal to eighteen plus seven times *b*.

Translate:

Four times the difference of b and six

$$4 \quad \cdot \quad b - 6$$

is equal to eighteen plus seven times b.

$$= \quad 18 \quad + \quad 7 \quad \cdot \quad b$$

Solve:

$$4(b-6) = 18 + 7b$$

$$4b - 24 = 18 + 7b \qquad \text{Distribute the 4.}$$

$$\underline{-4b \qquad\qquad -4b} \qquad \text{Subtract } 4b.$$

$$0 - 24 = 18 + 3b$$

$$-24 = 18 + 3b$$

$$\underline{-18 \quad -18} \qquad \text{Subtract 18.}$$

$$-42 = 0 + 3b$$

$$-42 = 3b$$

$$\dfrac{}{3} \quad \dfrac{}{3} \qquad \text{Divide by 3.}$$

$$-14 = b$$

3. Translate to an equation; then solve. Four more than the product of six and *n* is the same as four times the sum of *n* and twelve.

ADDITIONAL EXERCISES
Objective 1 Translate sentences to equations using key words; then solve.
For extra help, see Examples 1–3 on pages 247–248 of your text and the Section 4.4 lecture video.
Translate to an equation, then solve.

1. Fourteen less than a number is ten.

2. The sum of a number and eight is negative twelve.

3. A number increased by eleven is negative nineteen.

4. A number decreased by six is negative twenty.

5. The product of negative five and a number is thirty-five.

6. Negative one times a number is equal to twenty-six.

Name:
Instructor:
Date:
Section:

7. A number multiplied by eight is negative sixteen.

8. Negative three times a number is equal to negative eighteen.

9. The sum of eleven and four times t is the same as the difference of six times t and thirteen.

10. Six times x plus three times the difference of x and twenty is equal to sixty-two minus the sum of x and two.

11. Twenty less than negative ten times the difference of y and fifteen is the same as the product of negative four and y minus the difference of y and five.

12. Four times the sum of b and eight is equal to negative forty-eight.

13. Six times m subtracted from seven times the sum of m and three is equal to three times the sum of m and three.

Chapter 4 EQUATIONS

4.5 Applications and Problem Solving

Learning Objectives
1   Solve problems involving two unknown amounts.
2   Use a table in solving problems with two unknown amounts.

Key Terms

Use the vocabulary terms listed below to complete each statement in Exercises 1–4.

equilateral	isosceles	congruent	similar
complementary	supplementary	perimeter	area

1. Two angles whose sum is 90° are _____ angles.

2. _____ angles have a sum of 180°.

3. A triangle with sides of lengths 3 cm, 3 cm, and 5 cm is a(n) _____ triangle.

4. Angles that have the same measurement are _____ angles.

GUIDED EXAMPLES AND PRACTICE

Objective 1 Solve problems involving two unknown amounts.

Review these examples for Objective 1:

1. The perimeter of a college basketball court is 96 meters and the length is twice as long as the width. What are the length and width?

$$P = 2l + 2w$$ Perimeter is 2 times length plus 2 times width.

$$l = 2w$$ Length is twice width.

$$96 = 2(2w) + 2w$$ Replace P with 96 and l with $2w$.

$$96 = 6w$$ Combine like terms.

$$\frac{96}{6} = \frac{6w}{6}$$ Divide by 6.

$$16 = w$$

Practice these exercises:

1. Translate to an equation and solve. One number is six times another. The sum of the numbers is 266. What are the numbers?

Now solve for length.

$l = 2w$

$l = 2(16)$ Replace w with 16.

$l = 32$

Check:

$2 \cdot 16 + 2 \cdot 32 = 96$

Length is 32 meters and width is 16 meters.

2. A banner in the shape of an isosceles triangle has a base that is 6 inches shorter than either of the equal sides and the perimeter of the banner is 27 inches. What is the length of the equal sides?

Let s stand for the length of either of the two equal sides.

The base is $s - 6$.

$P = 2s + s - 6$ This is the equation

                        for the perimeter.

$27 = 2s + s - 6$ Replace P with 27.

$27 = 3s - 6$ Combine like terms.

$\underline{+6 \quad\quad +6}$ Add 6 to both sides.

$33 = 3s - 0$

$\dfrac{33}{3} = \dfrac{3s}{3}$ Divide by 3.

$11 = s$

Check:

$2 \cdot 11 + (11 - 6) = 27$

The length is 11 inches.

2. In the figure shown, each side of the equilateral triangle has the same measure as the length of the rectangle. The rectangle's width is 35 feet less than 4 times the length. The shapes have equal perimeters. Find the dimensions of each shape.

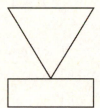

3. The sum of the measures of two complementary angles is 90°. If one angle measures 9° more than twice the measure of the other, find the measure of the smaller angle.

Let a stand for the larger angle and s stand for the smaller angle.

$a + s = 90$ The sum of complementary angles is 90°.

$9 + 2s = a$ One angle measures 9° more than twice the other.

Use these relationships to solve for s.

$(9 + 2s) + s = 90$ In the first equation, replace a with $9 + 2s$.

$$9 + 3s = 90$$
$$\underline{-9\qquad -9}\quad \text{Subtract 9 from both sides.}$$
$$0 + 3s = 81$$
$$\frac{3s}{3} = \frac{81}{3}\quad \text{Divide by 3.}$$
$$s = 27$$

The smaller angle measures 27°.

3. The sum of the measures of the angles of any triangle is 180°. In triangle ABC, angles A and B have the same measure, while the measure of angle C is 39° more than each of A and B. What are the measures of the three angles?

Objective 2 Use a table in solving problems with two unknown amounts.

Review this example for Objective 2:

4. Complete a four-column table; write an equation, and solve. A cashier has a total of 35 bills, made up of tens and fifties. The total value of the money is $1030. How many of each kind does he have?

Categories	Value	Number	Amount
	5		

Let b stand for the number of tens. Since the total number of bills is 35, then the number of fifties is $35 - b$.

Categories	Value	Number	Amount
Tens	10	b	$10b$
Fifties	50	$35 - b$	$50(35-b)$

Write the equation that represents the total value of money by using the value of the bills multiplied by the number of bills.

$$10b + 50(35 - b) = 1030$$

Solve the equation.

$$10b + 50(35 - b) = 1030$$

$10b + 1750 - 50b = 1030$ Distribute the 50.

$-40b + 1750 = 1030$ Combine like terms.

$\underline{-1030 \quad -1030}$ Subtract 1030.

$-40b + 720 = 0$

$\underline{+40b \qquad +40b}$ Add $40b$.

$0 + 720 = 40b$

$720 = 40b$

$\dfrac{720}{40} \quad \dfrac{}{40}$ Divide by 40.

$18 = b$

Practice this exercise:

4. Complete a four-column table; write an equation, and solve. A woman owns two stocks. Stock A is worth 20 times as much as stock B. She has 12 shares of stock A and 13 shares of stock B. The total worth of the stocks is $1265. Find each stock's share price.

Categories	Value	Number	Amount

Now solve for the number of fifties.

$35 - b =$ number of fifties

$35 - 18 = 17$ Replace b with 18

and solve.

The cashier has 18 tens and 17 fifties.

ADDITIONAL EXERCISES

Objective 1 Solve problems involving two unknown amounts.

For extra help, see Examples 1–4 on pages 251–255 of your text and the Section 4.5 lecture video.

Translate to an equation, then solve.

1. One number is 5 more than a second number. The sum of the numbers is twenty-nine. Find the numbers.

2. A decorator uses 66 feet of wallpaper border in a room. The length of the room is 5 feet more than the width. What are the dimensions?

3. In the figure shown, each side of the equilateral triangle has the same measure as the length of the rectangle. The rectangle's width is 30 feet less than 3 times the length. The shapes have equal perimeters. Find the dimensions of each shape.

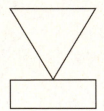

4. The sum of the angle measurements in any triangle is $180°$. One of the angles measures $55°$; the measure of the second is five more than three times the measure of the third. Find the two unknown angle measurements.

Objective 2 Use a table in solving problems with two unknown amounts.

For extra help, see Examples 5–6 on pages 257–258 of your text and the Section 4.5 lecture video.

Complete a four-column table; write an equation, and solve.

5.   A pair of pants costs $6 more than a knit shirt. The total cost to purchase 5 knit shirts and 3 pairs of pants is $170. Find the price of each item.

Categories	Value	Number	Amount

6.   A family of two adults and three children fly from Connecticut to Florida for a vacation. Each adult airline tickets costs $85 more than each child ticket. The total cost for the flights is $1345. Find the cost of each ticket.

Categories	Value	Number	Amount

7.   An animal shelter has dogs and cats available for adoption. The shelter charges $75 more to adopt a dog than to adopt a cat. One day the shelter had $1725 in income from pet adoptions. If 9 cats and 6 dogs were adopted that day, how much does the shelter charge to adopt each animal?

Categories	Value	Number	Amount

8.   A teacher purchased notebooks and boxes of pencils for her students. Each notebook cost three times as much as each box of pencils. The teacher has 21 students, and she bought a notebook and box of pencils for each student in her class. If she spent $84 on these supplies, how much did each item cost?

Categories	Value	Number	Amount

Chapter 5 FRACTIONS AND RATIONAL EXPRESSIONS

5.1 Introduction to Fractions

Learning Objectives
1   Name a fraction represented by a shaded region.
2   Write improper fractions as mixed numbers.
3   Write mixed numbers as improper fractions.
4   Graph fractions or mixed numbers on a number line.
5   Write equivalent fractions.
6   Use <, >, or = to write a true statement.

Key Terms
Use the vocabulary terms listed below to complete each statement in Exercises 1–5.

fraction	numerator	rational	simplify	improper
equivalent	greatest	multiple	dividend	mixed
denominator	fewest	simplest form		

1. An expression is in simplest form if it is written with the _____ symbols possible.

2. In a fraction, the number written in the top position is the _____.

3. If a number is evenly divisible by a given number, that number is a _____ of the given number.

4. A(n) _____ number can be written as a ratio of integers.

5. An integer combined with a fraction is a(n) _____ number.

GUIDED EXAMPLES AND PRACTICE

Objective 1 Name a fraction represented by a shaded region.

Review these examples for Objective 1:
1. Name the fraction represented by the shaded region.

Practice these exercises:
1. Name the fraction represented by the shaded region.

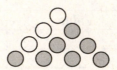

The figure has been divided into eight regions of equal size. Seven regions have been shaded, so the fraction $\frac{7}{8}$ describes the shaded region.

2. In an American Sign Language (A.S.L.) class of 19 students, 5 are hearing impaired. What fraction of the students are hearing impaired?

The fraction of the students that are hearing impaired is $\frac{5}{19}$.

2. Of 61 cars making up a freight train, 30 are boxcars. What fraction of the cars are not boxcars?

Objective 2 Write improper fractions as mixed numbers.

Review these examples for Objective 2:

3. Simplify $\frac{-31}{-31}$.

$$\frac{-31}{-31} = 1$$

4. Write $\frac{10}{7}$ as a mixed number.

$$7\overline{)10}$$
$$\underline{-7}$$
$$3$$

$$\frac{10}{7} = 1\frac{3}{7}$$

The quotient becomes the integer, the remainder becomes the new numerator, and the denominator stays the same.

5. Divide $23 \div 6$ and write the quotient as a mixed number.

$$6\overline{)23}$$
$$\underline{-18}$$
$$5$$

$$23 \div 6 = 3\frac{5}{6}$$

The quotient becomes the integer, the remainder becomes the new numerator, and the denominator stays the same.

Practice these exercises:

3. Simplify $\frac{0}{7}$.

4. Write $\frac{23}{4}$ as a mixed number.

5. Divide $-61 \div 7$ and write the quotient as a mixed number.

Objective 3 Write mixed numbers as improper fractions.

Review this example for Objective 3:	Practice this exercise:

6. Write $6\dfrac{7}{8}$ as an improper fraction.

6. Write $-10\dfrac{2}{3}$ as an improper fraction.

$$6\frac{7}{8} = \frac{8 \cdot 6 + 7}{8} \quad \text{Rewrite the numerator.}$$

$$= \frac{48 + 7}{8} \quad \text{Multiply.}$$

$$= \frac{55}{8} \quad \text{Add.}$$

Objective 4 Graph fractions or mixed numbers on a number line.

Review these examples for Objective 4:	Practice these exercises:

7. Graph $\dfrac{3}{10}$ on a number line.

7. Graph $-\dfrac{1}{3}$ on a number line.

The denominator is 10; so draw 10 evenly spaced marks to divide the segment between 0 and 1 into 10 equal spaces. The numerator is 3; so draw a dot on the third mark to the right of 0.

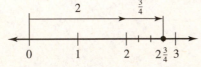

8. Graph $2\dfrac{3}{4}$ on a number line.

8. Graph $-1\dfrac{1}{3}$ on a number line.

The denominator is 4; so draw 4 evenly spaced marks to divide the segment between 2 and 3 into 4 equal spaces. The numerator is 3; so draw a dot on the third mark to the right of 2.

Name: Date:
Instructor: Section:

Objective 5 Write equivalent fractions.

Review these examples for Objective 5:

9.   Fill in the blank so that the fractions
     are equivalent.

$$\frac{9}{10} = \frac{?}{20}$$

$\frac{9}{10} = \frac{9 \cdot 2}{10 \cdot 2}$ Multiplying 10 by 2 produces
20; so multiplying 9 by 2 will
produce the unknown numerator.

$= \frac{18}{20}$ Multiply the numerator and the
denominator by 2.

10. Fill in the blank so that the fractions
     are equivalent.

$$\frac{16}{40} = \frac{2}{?}$$

$\frac{16}{40} = \frac{16 \div 8}{40 \div 8}$ Dividing 16 by 8 gives 2; so
dividing 40 by 8 will give the
unknown denominator.

$= \frac{2}{5}$ Divide the numerator and
denominator by 8.

Practice these exercises:

9.   Fill in the blank so that the
     fractions are equivalent.

$$\frac{-7}{10} = \frac{?}{40}$$

10. Fill in the blank so that the
     fractions are equivalent.

$$\frac{-8}{12} = \frac{-2}{?}$$

Objective 6 Use <, >, or = to write a true statement.

Review this example for Objective 3:

11. Use <, >, or = to write a true
     statement.

$$\frac{1}{3} \; ? \; \frac{6}{25}$$

Notice that a common multiple of 3
and 25 is 75, which can be found by
multiplying 25 times 3.

Practice this exercises:

11. Use <, >, or = to write a true
     statement.

$$\frac{3}{5} \; ? \; \frac{5}{7}$$

To write $\dfrac{1}{3}$ with a denominator of 75,
multiply its numerator and
denominator by 25.

$$\dfrac{1}{3} = \dfrac{1 \cdot 25}{3 \cdot 25} = \dfrac{25}{75}$$

To write $\dfrac{6}{25}$ with a denominator of
75, multiply its numerator and
denominator by 3.

$$\dfrac{6}{25} = \dfrac{6 \cdot 3}{25 \cdot 3} = \dfrac{18}{75}$$

$$\dfrac{25}{75} > \dfrac{18}{75}, \text{ so } \dfrac{1}{3} > \dfrac{6}{25}$$

ADDITIONAL EXERCISES
Objective 1 Name a fraction represented by a shaded region.
For extra help, see Examples 1–2 on pages 273–274 of your text and the Section 5.1 lecture video.

Name a fraction represented by the shaded region.

1.

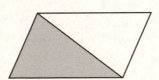

2.

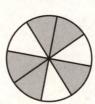

Write the fraction for each situation.

3. Twelve students board a school bus in the morning, and 11 students return on the bus in the afternoon. What fraction of the students that board the bus in the morning do not return on the bus in the afternoon?

4. Four of 13 friends do not have siblings. What fraction of the friends have siblings?

Objective 2 Write improper fractions as mixed numbers.
For extra help, see Examples 3–5 on pages 274–276 of your text and the Section 5.1 lecture video.
Simplify.

5. $\dfrac{-17}{1}$

6. $\dfrac{960}{0}$

Write the improper fraction as a mixed number.

7. $\dfrac{47}{6}$

8. $-\dfrac{75}{8}$

Divide and write the quotient as a mixed number.

9. $179 \div 14$

10. $-59 \div 4$

Objective 3 Write mixed numbers as improper fractions.
For extra help, see Example 6 on page 277 of your text and the Section 5.1 lecture video.
Write the mixed number as an improper fraction.

11. $-2\dfrac{3}{7}$

12. $6\dfrac{3}{5}$

Objective 4 Graph fractions or mixed numbers on a number line.
For extra help, see Example 7 on page 278 of your text and the Section 5.1 lecture video.
Graph the fraction on a number line.

13. $\dfrac{4}{5}$

14. $\dfrac{-3}{4}$

15. $3\dfrac{1}{2}$

16. $-4\dfrac{3}{4}$

Objective 5 Write equivalent fractions.
For extra help, see Example 8 on page 279 of your text and the Section 5.1 lecture video.
Fill in the blank so that the fractions are equivalent.

17. $\dfrac{5}{7} = \dfrac{?}{56}$ 18. $\dfrac{54}{21} = \dfrac{18}{?}$

Objective 6 Use <, >, or = to write a true statement.
For extra help, see Example 9 on pages 280–281 of your text and the Section 5.1 lecture video.
Use <, >, or = to write a true statement.

19. $\dfrac{2}{4} \; ? \; \dfrac{4}{5}$ 20. $-\dfrac{1}{5} \; ? \; -\dfrac{2}{10}$

21. $-\dfrac{2}{3} \; ? \; -\dfrac{11}{15}$ 22. $\dfrac{4}{6} \; ? \; \dfrac{10}{15}$

Chapter 5 FRACTIONS AND RATIONAL EXPRESSIONS

5.2 Simplifying Fractions and Rational Expressions

Learning Objectives
1   Simplify fractions to lowest terms.
2   Simplify improper fractions or fractions within mixed numbers.
3   Simplify rational expressions.

Key Terms

Use the vocabulary terms listed below to complete each statement in Exercises 1–4.

lowest   greatest   numerator   denominator   least   rational

1. A _____ expression is a ratio of monomials or polynomials.

2. A fraction is in _____ terms if the greatest common factor of its numerator and denominator is 1.

3. To simplify a fraction to lowest terms, divide the numerator and denominator by their _____ common factor.

4. To begin simplifying a fraction to lowest terms, replace the _____ and _____ with their prime factorizations.

GUIDED EXAMPLES AND PRACTICE

Objective 1 Simplify fractions to lowest terms.

Review these examples for Objective 1:

1. Simplify $\dfrac{24}{42}$ to lowest terms.

$$\frac{24}{42} = \frac{24 \div 6}{42 \div 6} \qquad \text{Divide both 24 and 42 by their GCF, 6.}$$

$$= \frac{4}{7} \qquad \text{Simplify.}$$

2. Simplify $-\dfrac{48}{72}$ to lowest terms.

The prime factorization of 48 is $2 \cdot 2 \cdot 2 \cdot 2 \cdot 3$.
The prime factorization of 72 is $2 \cdot 2 \cdot 2 \cdot 3 \cdot 3$.

Practice these exercises:

1. Simplify $\dfrac{21}{35}$ to lowest terms.

2. Simplify $-\dfrac{96}{120}$ to lowest terms.

$$-\frac{48}{72} = -\frac{\overset{1}{\cancel{2}}\cdot\overset{1}{\cancel{2}}\cdot\overset{1}{\cancel{2}}\cdot 2 \cdot \overset{1}{\cancel{3}}}{\underset{1}{\cancel{2}}\cdot\underset{1}{\cancel{2}}\cdot\underset{1}{\cancel{2}}\cdot\underset{1}{\cancel{3}}\cdot 3}$$

Dividing out the common factors divides out the GCF.

$$= -\frac{2}{3} \qquad \text{Multiply.}$$

3. A 30-foot picket fence is being painted white. So far, 18 feet of the fence has been painted. What is the fraction, in lowest terms, of how much of the fence is not painted yet?

If 18 feet of fence has been painted, then $30 - 18 = 12$ feet of the fence has not been painted.

$$\frac{12}{30} = \frac{12 \div 6}{30 \div 6} \quad \begin{array}{l}\text{Divide both 12 and 30 by} \\ \text{their GCF, 6.}\end{array}$$

$$= \frac{2}{5} \qquad \text{Simplify.}$$

3. A baseball player has hit 24 home runs so far this season. Nine of those home runs have come in games against division rivals. What is the fraction, in lowest terms, of the number of home runs not hit against division rivals?

Objective 2 Simplify improper fractions or fractions within mixed numbers.

Review these examples for Objective 2:

4. Write $\frac{10}{4}$ as a mixed number.

$$\frac{10}{4} = \frac{5 \cdot \overset{1}{\cancel{2}}}{2 \cdot \underset{1}{\cancel{2}}} = \frac{5}{2} \qquad \text{Simplify to lowest terms.}$$

$$= 2\frac{1}{2} \qquad \begin{array}{l}\text{Write the mixed number. Its} \\ \text{fraction is in lowest terms.}\end{array}$$

5. Write $\frac{63}{12}$ as a mixed number.

Practice these exercises:

4. Write $\frac{20}{6}$ as a mixed number.

5. Write $\frac{38}{8}$ as a mixed number.

$\dfrac{63}{12} = 5\dfrac{3}{12}$ Write as a mixed number.

$= 5\dfrac{\overset{1}{\cancel{3}}}{2 \cdot 2 \cdot \underset{1}{\cancel{3}}}$ Simplify the mixed number's fraction.

$= 5\dfrac{1}{4}$.

Objective 3 Simplify rational expressions.

Review these examples for Objective 3:

6.   Simplify $\dfrac{2n^4}{8n}$ to lowest terms.

$$\dfrac{2n^4}{8n} = \dfrac{2 \cdot n \cdot n \cdot n \cdot n}{2 \cdot 2 \cdot 2 \cdot n}$$ Write the numerator and denominator in factored form

$$= \dfrac{\overset{1}{\cancel{2}} \cdot \overset{1}{\cancel{n}} \cdot n \cdot n \cdot n}{\underset{1}{\cancel{2}} \cdot 2 \cdot 2 \cdot \underset{1}{\cancel{n}}}$$ Divide out a 2 and an n.

$$= \dfrac{n^3}{4}$$ Multiply the remaining factors.

7.   Simplify $\dfrac{12t^3u^3v^2}{18tv^5}$ to lowest terms.

$$\dfrac{12t^3u^3v^2}{18tv^5} = \dfrac{2 \cdot 2 \cdot 3 \cdot t \cdot t \cdot t \cdot u \cdot u \cdot u \cdot v \cdot v}{2 \cdot 3 \cdot 3 \cdot t \cdot u \cdot u \cdot u \cdot u \cdot u}$$

Write the numerator and denominator in factored form.

$$= \dfrac{\cancel{2} \cdot 2 \cdot \cancel{3} \cdot \cancel{t} \cdot t \cdot t \cdot u \cdot u \cdot u \cdot \cancel{v} \cdot \cancel{v}}{\cancel{2} \cdot \cancel{3} \cdot 3 \cdot \cancel{t} \cdot \cancel{v} \cdot \cancel{v} \cdot v \cdot v \cdot v}$$

Divide out one 2, one 3, one t, and two v's.

$$= \dfrac{2t^2u^3}{3v^3}$$

Multiply the remaining factors.

Practice these exercises:

6.   Simplify $\dfrac{6k^6}{9k^5}$ to lowest terms.

7.   Simplify $\dfrac{24b^5cd^4}{40b^3c^4d}$ to lowest terms.

Name: _____   Date: _____

Instructor: _____   Section: _____

ADDITIONAL EXERCISES
Objective 1 Simplify fractions to lowest terms.
For extra help, see Examples 1–3 on pages 285–287 of your text and the Section 5.2 lecture video.

Simplify to lowest terms.

1. $\dfrac{9}{15}$

2. $\dfrac{19}{57}$

3. $-\dfrac{28}{36}$

4. $\dfrac{48}{132}$

5. $\dfrac{60}{108}$

6. $\dfrac{205}{287}$

Write each requested fraction in lowest terms.

7. What fraction of a minute is 51 seconds?

8. What fraction of September is 18 days?

9. A survey is conducted in front of a home improvement store. The person conducting the survey asks 132 men if they use a certain cleansing agent and 120 respond that they do use the product. What is the fraction, in lowest terms, of the men surveyed that said they used the product?

10. A cubic meter of concrete mix contains 400 kilograms of cement, 160 kilograms of stone, and 110 kilograms of sand. What is the total weight of the cubic meter of concrete mix, and what part is stone?

Objective 2 Simplify improper fractions or fractions within mixed numbers.
For extra help, see Example 4 on page 287 of your text and the Section 5.2 lecture video.
Write each fraction as a mixed number and simplify.

11. $\dfrac{90}{55}$

12. $-\dfrac{85}{70}$

13. $-\dfrac{90}{65}$

14. $-\dfrac{111}{9}$

Objective 3 Simplify rational expressions.
For extra help, see Examples 5–6 on page 288 of your text and the Section 5.2 lecture video.
Simplify to lowest terms.

15. $\dfrac{18xy}{24x}$

16. $\dfrac{12j^2}{90jk}$

17. $\dfrac{-60c^3}{80c^2d}$

18. $\dfrac{10a^2}{490a^2b}$

19. $\dfrac{9m^2}{315m^2n}$

20. $\dfrac{30g^2i^3}{294ghi^2}$

Chapter 5 FRACTIONS AND RATIONAL EXPRESSIONS

5.3 Multiplying Fractions, Mixed Numbers, and Rational Expressions

Learning Objectives
1   Multiply fractions.
2   Multiply and simplify fractions.
3   Multiply mixed numbers.
4   Multiply rational expressions.
5   Simplify fractions raised to a power.
6   Solve applications involving multiplying fractions.
7   Find the area of a triangle.
8   Find the radius and diameter of a circle.
9   Find the circumference of a circle.

Key Terms
Use the vocabulary terms listed below to complete each statement in Exercises 1–4.

circle	center	radius	diameter
ratio	irrational	circumference	

1.   A number that cannot be expressed exactly as a fraction is a(n) _____
     number.

2.   A collection of points that are equally distant from a central point is called a(n)
     _____.

3.   The distance around a circle is its _____.

4.   The distance across a circle along a straight line through the center is called the
     _____.

GUIDED EXAMPLES AND PRACTICE

Objective 1   Multiply fractions.

Review these examples for Objective 1:

1.   Multiply $\dfrac{1}{3} \cdot \dfrac{2}{5}$.

     Multiply numerator by numerator and
     denominator by denominator.

     $$\frac{1}{3} \cdot \frac{2}{5} = \frac{1 \cdot 2}{3 \cdot 5} = \frac{2}{15}$$

Practice these exercises:

1.   Multiply $\dfrac{3}{4} \cdot \dfrac{3}{7}$.

2. Multiply $-\dfrac{1}{2} \cdot \dfrac{5}{9}$.

Multiply numerator by numerator and denominator by denominator. The product is negative because the two fractions have different signs.

$$-\dfrac{1}{2} \cdot \dfrac{5}{9} = -\dfrac{1 \cdot 5}{2 \cdot 9} = -\dfrac{5}{18}$$

3. Multiply $-\dfrac{5}{11} \cdot \left(-\dfrac{4}{13}\right)$.

Multiply numerator by numerator and denominator by denominator. The product is positive because the two fractions have the same signs.

$$-\dfrac{5}{11} \cdot \left(-\dfrac{4}{13}\right) = \dfrac{5 \cdot 4}{11 \cdot 13} = \dfrac{20}{143}$$

2. Multiply $\dfrac{2}{5} \cdot \left(-\dfrac{4}{7}\right)$.

3. Multiply $-\dfrac{3}{5} \cdot \left(-\dfrac{7}{8}\right)$.

Objective 2 Multiply and simplify fractions.

Review these examples for Objective 2:

4. Multiply $\dfrac{4}{6} \cdot \dfrac{9}{15}$. Write the product in lowest terms.

$$\dfrac{4}{6} \cdot \dfrac{9}{15} = \dfrac{36}{90} \qquad \text{Multiply.}$$

$$= \dfrac{\overset{1}{\cancel{2}} \cdot 2 \cdot \overset{1}{\cancel{3}} \cdot \overset{1}{\cancel{3}}}{\underset{1}{\cancel{2}} \cdot \underset{1}{\cancel{3}} \cdot \underset{1}{\cancel{3}} \cdot 5}$$

Simplify to lowest terms by dividing out the common factors.

$$= \dfrac{2}{5}$$

Practice these exercises:

4. Multiply $\dfrac{2}{9} \cdot \dfrac{3}{8}$. Write the product in lowest terms.

5. Multiply $\dfrac{10}{15} \cdot \dfrac{12}{30}$. Write the product in lowest terms.

$$\frac{10}{15} \cdot \frac{12}{30} = \frac{2 \cdot 5}{3 \cdot 5} \cdot \frac{2 \cdot 2 \cdot 3}{2 \cdot 3 \cdot 5}$$

Replace each numerator and denominator with its prime factorization.

$$= \frac{2 \cdot \overset{1}{\cancel{5}}}{3 \cdot \cancel{5}} \cdot \frac{\overset{1}{\cancel{2}} \cdot 2 \cdot \overset{1}{\cancel{3}}}{\underset{1}{\cancel{2}} \cdot \underset{1}{\cancel{3}} \cdot 5}$$

Divide out the common factors.

$$= \frac{4}{15}$$

5. Multiply $\dfrac{9}{21} \cdot \dfrac{14}{33}$. Write the product in lowest terms.

Objective 3 Multiply mixed numbers.

Review these examples for Objective 3:

6. Estimate $2\dfrac{2}{3} \cdot 4\dfrac{1}{5}$, then find the actual product expressed as a mixed number in lowest terms.

$2\dfrac{2}{3}$ rounds up to 3. $4\dfrac{1}{5}$ rounds down to 4. The actual product should be around $3 \cdot 4 = 12$.

$$2\frac{2}{3} \cdot 4\frac{1}{5} = \frac{8}{3} \cdot \frac{21}{5}$$ Write as improper fractions.

$$= \frac{2 \cdot 2 \cdot 2}{\underset{1}{\cancel{3}}} \cdot \frac{\overset{1}{\cancel{3}} \cdot 7}{5}$$ Divide out a common factor, 3.

$$= \frac{56}{5}$$ Multiply.

$$= 11\frac{1}{5}$$ Write as a mixed number.

The actual product agrees with the estimate.

Practice these exercises:

6. Estimate $4\dfrac{1}{2} \cdot 2\dfrac{1}{6}$, then find the actual product expressed as a mixed number in lowest terms.

7. Estimate $12 \cdot \dfrac{4}{9}$, then find the actual product expressed as a mixed number in lowest terms.

Rounding $\dfrac{4}{9}$ down to 0 is not helpful,

so instead round $\dfrac{4}{9}$ to $\dfrac{1}{2}$. The actual

product should be around $12 \cdot \dfrac{1}{2} = 6$.

$$12 \cdot \frac{4}{9} = \frac{12}{1} \cdot \frac{4}{9} \qquad \text{Write as improper fractions.}$$

$$= \frac{2 \cdot 2 \cdot \overset{1}{\cancel{3}}}{1} \cdot \frac{2 \cdot 2}{\underset{1}{\cancel{3}} \cdot 3} \qquad \text{Divide out a common factor, 3.}$$

$$= \frac{16}{3} \qquad \text{Multiply.}$$

$$= 5\frac{1}{3} \qquad \text{Write as a mixed number.}$$

The actual product agrees with the estimate.

8. Multiply $-4\dfrac{2}{9} \cdot 3\dfrac{3}{4}$. Write the product as a mixed number in lowest terms.

$$-4\frac{2}{9} \cdot 3\frac{3}{4} = -\frac{38}{9} \cdot \frac{15}{4}$$

Write as improper fractions.

$$= -\frac{19 \cdot \overset{1}{\cancel{2}}}{3 \cdot \underset{1}{\cancel{3}}} \cdot \frac{\overset{1}{\cancel{3}} \cdot 5}{\underset{1}{\cancel{2}} \cdot 2}$$

Divide out the common factors, which are 2 and 3.

7. Estimate $10 \cdot \dfrac{5}{8}$, then find the actual product expressed as a mixed number in lowest terms.

8. Multiply $2\dfrac{1}{2} \cdot 4\dfrac{3}{5}$. Write the product as a mixed number in lowest terms.

$$= -\frac{95}{6}$$

Multiply. The product is negative because the two numbers have different signs.

$$= -15\frac{5}{6}$$

Write the improper fractions as a mixed number.

Objective 4 Multiply rational expressions.

Review this example for Objective 4:

9. Multiply $\dfrac{4s^2}{7t} \cdot \dfrac{35st^3}{6}$. Write the product in lowest terms.

$$\frac{4s^2}{7t} \cdot \frac{35st^3}{6} = \frac{2 \cdot 2 \cdot s \cdot s}{7 \cdot t} \cdot \frac{5 \cdot 7 \cdot s \cdot t \cdot t \cdot t}{2 \cdot 3}$$

Write the prime factorization of each numerator and denominator.

$$= \frac{2 \cdot \overset{1}{\cancel{2}} \cdot s \cdot s}{\underset{1}{\cancel{7}} \cdot \underset{1}{\cancel{t}}} \cdot \frac{5 \cdot \overset{1}{\cancel{7}} \cdot s \cdot \overset{1}{\cancel{t}} \cdot t \cdot t}{\underset{1}{\cancel{2}} \cdot 3}$$

Divide out a common 2, a common 7, and a common t.

$$= \frac{10s^3t^2}{3}$$

Multiply the remaining numerator and denominator factors.

Objective 5 Simplify fractions raised to a power.

Review these examples for Objective 5:

10. Simplify $\left(\dfrac{3}{5}\right)^3$.

Practice this exercise:

9. Multiply $-\dfrac{3xy}{8} \cdot \left(-\dfrac{10}{9x^3y}\right)$. Write the product in lowest terms.

Practice these exercises:

10. Simplify $\left(\dfrac{2}{9}\right)^4$.

Write the base $\frac{3}{5}$ as a factor three

times; then multiply.

$$\left(\frac{3}{5}\right)^3 = \frac{3}{5} \cdot \frac{3}{5} \cdot \frac{3}{5}$$

$$= \frac{27}{125}$$

11. Simplify $\left(\dfrac{-4pr^3}{7q^2}\right)^3$.

$$\left(\frac{-4pr^3}{7q^2}\right)^3 = \frac{\left(-4pr^3\right)^3}{\left(7q^2\right)^3}$$

Write both the numerator and

denominator raised to the third power.

$$= \frac{\left(-4\right)^3 p^3 r^{3\cdot3}}{7^3 q^{2\cdot3}}$$

Write the coefficient to the 3rd power and

multiply each variable's exponent by that

power.

$$= \frac{-64 p^3 r^9}{343 q^6} \qquad \text{Simplify.}$$

11. Simplify $\left(\dfrac{9h^4}{-8j^2k}\right)^2$.

Objective 6 Solve applications involving multiplying fractions.

Review these examples for Objective 6:

12. Three of every 7 pies at a bake sale are apple pies. If there are 84 pies at the bake sale, how many are apple pies?

3 of every 7 is the fraction $\frac{3}{7}$. This

means $\frac{3}{7}$ of the 84 pies are apple.

Practice these exercises:

12. A mayoral candidate is confident he will win 7 of every 12 votes in his town. If 372 people vote in the first two hours after polls open, how many of those votes does the candidate expect he will get?

$$\frac{3}{7} \cdot 84 = \frac{3}{7} \cdot \frac{84}{1}$$ Write as improper fractions.

$$= \frac{3}{\overset{}{\underset{1}{\cancel{7}}}} \cdot \frac{2 \cdot 2 \cdot 3 \cdot \overset{1}{\cancel{7}}}{1}$$ Divide out a common factor, 7.

$$= 36$$ Multiply.

There are 36 apple pies.

13. A billiards trick shot artist makes $\frac{5}{6}$ of her shots in competition. Of those shots, she makes $\frac{3}{5}$ on her first attempt. What fraction of her shots does she make on her first attempt?

$\frac{3}{5}$ of the $\frac{5}{6}$ successful shots were made on the artist's first attempt.

$$\frac{3}{5} \cdot \frac{5}{6} = \frac{\overset{1}{\cancel{3}}}{\cancel{5}} \cdot \frac{\overset{1}{\cancel{5}}}{2 \cdot \underset{1}{\cancel{3}}}$$ Divide out common factors.

$$= \frac{1}{2}$$ Multiply.

She makes $\frac{1}{2}$ of her shots on her first attempt.

13. A group of elementary school students are at a lakeside summer camp. Nine-sixteenths of the group go swimming in a part of the lake designated for swimming. Of the swimmers, $\frac{2}{3}$ pass the test allowing them to swim in the deeper parts of the designated swimming area. What fraction of the group of students are allowed to swim in the deeper parts of the designated swimming area?

Objective 7 Find the area of a triangle.

Review this example for Objective 7:

14. A child's triangular building blocks are $3\frac{2}{5}$ cm along the base and $2\frac{1}{3}$ cm high. Find the area of the triangular face of the block.

Practice this exercise:

14. Find the area of a triangle with height $7\frac{1}{2}$ in. and base $2\frac{1}{2}$ in.

$$A = \frac{1}{2}bh = \frac{1}{2}\left(3\frac{2}{5}\right)\left(2\frac{1}{3}\right)$$

Replace b with $3\frac{2}{5}$ and h with $2\frac{1}{3}$.

$$= \frac{1}{2}\left(\frac{17}{5}\right)\left(\frac{7}{3}\right)$$

Write the mixed numbers as improper fractions.

$$= \frac{119}{30} \quad \text{Multiply.}$$

$$= 3\frac{29}{30} \quad \begin{array}{l}\text{Write the result as an} \\ \text{improper fraction.}\end{array}$$

The area is $3\frac{29}{30}$ cm^2.

Objective 8 Find the radius and diameter of a circle.

Review these examples for Objective 8:

15. Find the diameter of a circle with radius $1\frac{7}{8}$ ft.

$$d = 2r = 2\left(1\frac{7}{8}\right) \quad \text{Replace } r \text{ with } 1\frac{7}{8}$$

$$= \frac{\overset{1}{\cancel{2}}}{1}\cdot\frac{15}{\underset{4}{\cancel{8}}}$$

Write the mixed number as an improper fraction. Then divide out the common factor, 2.

$$= \frac{15}{4} \quad \text{Multiply.}$$

$$= 3\frac{3}{4} \quad \begin{array}{l}\text{Write the result as a} \\ \text{mixed number.}\end{array}$$

The diameter is $3\frac{3}{4}$ ft.

Practice these exercises:

15. Find the diameter of a circle with radius $4\frac{5}{6}$ in.

16. Find the radius of a circle with diameter 13 cm.

$$r = \frac{1}{2}d = \frac{1}{2}(13) \quad \text{Replace } d \text{ with 13.}$$

$$= \frac{1}{2} \cdot \frac{13}{1} \quad \text{Write as an improper fraction.}$$

$$= \frac{13}{2} \quad \text{Multiply.}$$

$$= 6\frac{1}{2} \quad \text{Write as a mixed number.}$$

The radius is $6\frac{1}{2}$ cm.

16. Find the radius of a circle with diameter $3\frac{1}{7}$ m.

Objective 9   Find the circumference of a circle.

Review this example for Objective 9:

17. Find the circumference of a circle with a radius of $3\frac{3}{8}$ in. Use $\frac{22}{7}$ for π.

$$C = 2\pi r = 2\left(\frac{22}{7}\right)\left(3\frac{3}{8}\right)$$

Replace π with $\frac{22}{7}$ and r with $3\frac{3}{8}$.

$$= \frac{\overset{1}{\cancel{2}}}{1} \cdot \frac{\overset{1}{\cancel{2}} \cdot 11}{7} \cdot \frac{3 \cdot 3 \cdot 3}{\cancel{2} \cdot \cancel{2} \cdot 2}$$

Write as improper fractions and divide out the common factors.

$$= \frac{297}{14} \quad \text{Multiply.}$$

$$= 21\frac{3}{14} \quad \text{Write as a mixed number.}$$

The circumference is $21\frac{3}{14}$ in .

Practice this exercise:

17. Find the circumference of a circle with a radius of $2\frac{5}{8}$ ft. Use $\frac{22}{7}$ for π.

ADDITIONAL EXERCISES
Objective 1 Multiply fractions.
For extra help, see Example 1 on page 292 of your text and the Section 5.3 lecture video.
Multiply.

1. $-\dfrac{1}{8} \cdot \dfrac{1}{6}$

2. $\dfrac{5}{7} \cdot \dfrac{6}{13}$

3. $\dfrac{3}{10} \cdot \dfrac{3}{100}$

4. $-\dfrac{7}{9} \cdot \left(-\dfrac{2}{9}\right)$

Objective 2 Multiply and simplify fractions.
For extra help, see Example 2 on page 293 of your text and the Section 5.3 lecture video.
Multiply. Write the product in lowest terms.

5. $\dfrac{6}{7} \cdot \dfrac{5}{9}$

6. $\dfrac{14}{39} \cdot \dfrac{3}{4}$

7. $-\dfrac{10}{39} \cdot \dfrac{3}{8}$

8. $\dfrac{16}{51} \cdot \dfrac{3}{10}$

9. $-\dfrac{16}{27} \cdot \left(-\dfrac{9}{32}\right)$

10. $\dfrac{-7}{15} \cdot \left(\dfrac{10}{-14}\right)$

Objective 3 Multiply mixed numbers.
For extra help, see Examples 3–4 on pages 294–295 of your text and the Section 5.3 lecture video.
Estimate and then find the actual product expressed as a mixed number in simplest form.

11. $3\dfrac{2}{5} \cdot 5\dfrac{3}{4}$

12. $45\left(\dfrac{4}{5}\right)$

Multiply. Write the product as a mixed number in simplest form.

13. $2\dfrac{3}{8} \cdot 1\dfrac{5}{8}$

14. $21 \cdot \dfrac{2}{3}$

Objective 4 Multiply rational expressions.
For extra help, see Example 5 on page 295 of your text and the Section 5.3 lecture video.
Multiply and write the product in lowest terms.

15. $\dfrac{z^2}{4} \cdot \dfrac{3}{5}$

16. $-\dfrac{y^2}{3} \cdot \dfrac{6}{5}$

17. $-\dfrac{x^4}{3yz} \cdot \dfrac{9y^3}{x^7 z}$

18. $-\dfrac{x^4}{5yz} \cdot \dfrac{25y^5}{x^6 z}$

Objective 5 Simplify fractions raised to a power.
For extra help, see Examples 6–7 on page 296 of your text and the Section 5.3 lecture video.
Simplify.

19. $\left(\dfrac{5}{4}\right)^2$

20. $\left(-\dfrac{2}{7}\right)^3$

21. $\left(\dfrac{w^9}{4}\right)^4$

22. $\left(\dfrac{-2m^2 n^7}{3p^9}\right)^3$

Objective 6 Solve applications involving multiplying fractions.
For extra help, see Examples 8–9 on pages 297–298 of your text and the Section 5.3 lecture video.
Solve. Write all answers in lowest terms.

23. Cheryl receives $32 for working a full day doing inventory at a hardware store. How much can she get for working $\dfrac{3}{8}$ of a day?

24. A recipe asks for $\dfrac{1}{7}$ cup of granola. How much is needed to make $\dfrac{6}{7}$ of the recipe?

Objective 7 Find the area of a triangle.
For extra help, see Example 10 on page 299 of your text and the Section 5.3 lecture video.
Find the area of each triangle.

25.

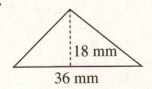

18 mm

36 mm

26.

$7\dfrac{1}{2}$ m

12 m

Objective 8 Find the radius and diameter of a circle.

For extra help, see Examples 11–12 on pages 300–301 of your text and the Section 5.3 lecture video.

Solve.

27. The radius of a circle is $\dfrac{5}{8}$ inch. What is its diameter?

28. The diameter of a circle is 15 centimeters. What is its radius?

Objective 9 Find the circumference of a circle.

For extra help, see Example 13 on page 302 of your text and the Section 5.3 lecture video.

Solve. Use $\dfrac{22}{7}$ for π.

29. The radius of a circle is $\dfrac{7}{10}$ inch. What is its circumference?

30. The radius of a circle is $1\dfrac{19}{44}$ m. What is its circumference?

Chapter 5 FRACTIONS AND RATIONAL EXPRESSIONS

5.4 Dividing Fractions, Mixed Numbers, and Rational Expressions

Learning Objectives
1   Divide fractions.
2   Divide mixed numbers.
3   Divide rational expressions.
4   Find the square root of a fraction.
5   Solve equations involving fractions.
6   Solve applications involving division of fractions.

Key Terms
Use the vocabulary terms listed below to complete each statement in Exercises 1–4.

reciprocal   complex·   numerator   denominator   improper

1.   A _____ fraction is a fractional expression with a fraction in its numerator and/or denominator.

2.   The _____ of a fraction is the same things as its multiplicative inverse.

3.   To simplify before multiplying fractions, divide out any numerator factors with any like _____ factors.

4.   To divide mixed numbers, begin by writing the mixed numbers as _____ fractions.

GUIDED EXAMPLES AND PRACTICE

Objective 1 Divide fractions.

Review these examples for Objective 1:

1.   Find the reciprocal of $\dfrac{2}{13}$.

     Invert the numerator and the denominator to get $\dfrac{13}{2}$.

     $$\dfrac{2}{13} \cdot \dfrac{13}{2} = 1$$

Practice these exercises:
1.   Find the reciprocal of 4.

2. Divide $\dfrac{8}{15} \div \dfrac{4}{5}$. Write the quotient in lowest terms.

$\dfrac{8}{15} \div \dfrac{4}{5} = \dfrac{8}{15} \cdot \dfrac{5}{4}$ Write an equivalent multiplication statement

$= \dfrac{2 \cdot \overset{1}{\cancel{2}} \cdot \overset{1}{\cancel{2}}}{3 \cdot \underset{1}{\cancel{5}}} \cdot \dfrac{\overset{1}{\cancel{5}}}{\underset{1}{\cancel{2}} \cdot \underset{1}{\cancel{2}}}$ Divide out the common factors.

$= \dfrac{2}{3}$ Multiply.

2. Divide $-\dfrac{6}{11} \div 4$. Write the quotient in lowest terms.

Objective 2 Divide mixed numbers.

Review this example for Objective 2:

3. Estimate $5\dfrac{3}{5} \div 2\dfrac{1}{3}$ and then find the actual quotient expressed as a mixed number in simplest form.

$5\dfrac{3}{5}$ rounds up to 6, and $2\dfrac{1}{3}$ rounds down to 2. The actual quotient should be around $6 \div 2 = 3$.

$5\dfrac{3}{5} \div 2\dfrac{1}{3} = \dfrac{28}{5} \div \dfrac{7}{3}$ Write the mixed numbers as improper fractions.

$= \dfrac{28}{5} \cdot \dfrac{3}{7}$ Write an equivalent multiplication statement.

$= \dfrac{2 \cdot 2 \cdot \overset{1}{\cancel{7}}}{5} \cdot \dfrac{3}{\underset{1}{\cancel{7}}}$ Divide out the common factor, 7.

$= \dfrac{12}{5}$ Multiply.

$= 2\dfrac{2}{5}$ Write as a mixed number.

Practice this exercise:

3. Estimate $7\dfrac{1}{3} \div \dfrac{4}{9}$ and then find the actual quotient expressed as a mixed number in simplest form.

Objective 3   Divide rational expressions.

Review this example for Objective 3:

4.   Divide $\dfrac{9cd^3}{20b^2} \div \dfrac{12c^2}{35b^4}$. Write the

quotient in lowest terms.

$$\frac{9cd^3}{20b^2} \div \frac{12c^2}{35b^4} = \frac{9cd^3}{20b^2} \cdot \frac{35b^4}{12c^2}$$

Write an equivalent multiplication statement.

$$= \frac{3 \cdot \cancel{3} \cdot \cancel{c} \cdot d \cdot d \cdot d}{2 \cdot 2 \cdot \cancel{5} \cdot \cancel{b} \cdot \cancel{b}} \cdot \frac{\cancel{5} \cdot 7 \cdot \cancel{b} \cdot \cancel{b} \cdot b \cdot b}{2 \cdot 2 \cdot \cancel{3} \cdot \cancel{c} \cdot c}$$

Write in factored form and divide out the

common factors.

$$= \frac{21b^2 d^3}{16c}$$

Multiply the remaining factors.

Practice this exercise:

4.   Divide $-\dfrac{10p}{7q^4} \div \left(-\dfrac{15p^2 q^2}{28r^3} \right)$. Write

the quotient in lowest terms.

Objective 4   Find the square root of a fraction.

Review these examples for Objective 4:

5.   Simplify $\sqrt{\dfrac{9}{4}}$.

Find the square root of the numerator and the denominator separately; then simplify.

$$\sqrt{\frac{9}{4}} = \frac{\sqrt{9}}{\sqrt{4}} = \frac{3}{2}$$

6.   Simplify $\sqrt{\dfrac{28}{7}}$.

Simplify the fraction; then find the square root of the quotient.

$$\sqrt{\frac{28}{7}} = \sqrt{4} = 2$$

Practice these exercises:

5.   Simplify $\sqrt{\dfrac{16}{81}}$.

6.   Simplify $\sqrt{\dfrac{48}{3}}$.

Objective 5 Solve equations involving fractions.

Review these examples for Objective 5:	Practice these exercises:
7. Solve and check. $\dfrac{5}{6}x = 2\dfrac{2}{9}$	7. Solve and check. $\dfrac{4}{7}y = 1\dfrac{3}{5}$

$\dfrac{6}{5} \cdot \dfrac{5}{6}x = \dfrac{20}{9} \cdot \dfrac{6}{5}$ Write the mixed number as an improper fraction, then multiply both sides by $\dfrac{6}{5}$.

$\dfrac{\cancel{6}}{\cancel{5}} \cdot \dfrac{\cancel{5}}{\cancel{6}}x = \dfrac{2 \cdot 2 \cdot \cancel{5}}{3 \cdot \cancel{3}} \cdot \dfrac{2 \cdot \cancel{3}}{\cancel{5}}$ Divide out the common factors.

$1x = \dfrac{8}{3}$ Multiply.

$x = \dfrac{8}{3}$, or $2\dfrac{2}{3}$

Check: $\dfrac{5}{6}x = 2\dfrac{2}{9}$

$\dfrac{5}{6} \cdot \dfrac{8}{3} \overset{?}{=} 2\dfrac{2}{9}$ Replace x with $\dfrac{8}{3}$ and verify that the equation is true.

$\dfrac{40}{18} \overset{?}{=} 2\dfrac{2}{9}$ Multiply.

$\dfrac{20}{9} \overset{?}{=} 2\dfrac{2}{9}$ Simplify.

$2\dfrac{2}{9} = 2\dfrac{2}{9}$ True; so $\dfrac{8}{3}$ is the solution.

| 8. Solve and check. $\dfrac{j}{4} = -2\dfrac{3}{8}$ | 8. Solve and check. $5\dfrac{2}{9} = \dfrac{k}{6}$ |

$$\frac{4}{1} \cdot \frac{j}{4} = -\frac{19}{8} \cdot \frac{4}{1}$$

Write the mixed number as an improper fraction, then multiply both sides by $\frac{4}{1}$.

$$\frac{\cancel{4}}{1} \cdot \frac{j}{\cancel{4}} = -\frac{19}{2 \cdot \cancel{4}} \cdot \frac{\cancel{4}}{1}$$ Divide out the common factors.

$$j = -\frac{19}{2}, \text{ or } -9\frac{1}{2}$$ Multiply.

Check: $\dfrac{j}{4} = -2\dfrac{3}{8}$

Replace j with $-\dfrac{19}{2}$ and verify that the equation is true.

$$\frac{-\dfrac{19}{2}}{4} \overset{?}{=} -2\frac{3}{8}$$

$$-\frac{19}{2} \div 4 \overset{?}{=} -2\frac{3}{8}$$

$$-\frac{19}{2} \cdot \frac{1}{4} \overset{?}{=} -2\frac{3}{8}$$ Write an equivalent multiplication statement.

$$-\frac{19}{8} \overset{?}{=} -2\frac{3}{8}$$ Multiply.

$$-2\frac{3}{8} = -2\frac{3}{8}$$ True; so $-\dfrac{19}{2}$ is the solution.

Objective 6 Solve applications using division of fractions.

Review this example for Objective 4:

9. A breakfast cereal box contains 49 servings of cereal. A certain bowl can hold $2\frac{1}{3}$ servings of cereal. How many bowls of cereal does the box contain?

Practice this exercise:

9. The circumference of a circle is $10\frac{3}{14}$ ft. What is the radius of the circle? Use $\frac{22}{7}$ for π.

$$2\frac{1}{3} \cdot b = 49$$

$$\frac{3}{7} \cdot \frac{7}{3} \cdot b = \frac{49}{1} \cdot \frac{3}{7}$$

Write the mixed numbers as improper fractions,

then multiply both sides by $\frac{3}{7}$.

$$\frac{\cancel{3}}{\cancel{7}} \cdot \frac{\cancel{7}}{\cancel{3}} \cdot b = \frac{7 \cdot \cancel{7}}{1} \cdot \frac{3}{7}$$ Divide out the common

                              factors.

$$b = 21$$ Multiply.

The box contains 21 bowls of cereal.

ADDITIONAL EXERCISES
Objective 1 Divide fractions.
For extra help, see Examples 1–2 on pages 307–308 of your text and the Section 5.4 lecture video.

Find the reciprocal.

1. $\dfrac{4}{9}$
                                   2. $\dfrac{1}{8}$

3. -11
                                   4. $-\dfrac{4}{7}$

Divide. Write the quotient in lowest terms

5. $\dfrac{1}{5} \div \dfrac{1}{15}$
                         6. $\dfrac{49}{45} \div \dfrac{7}{5}$

7. $-\dfrac{18}{13} \div 6$
                         8. $\dfrac{12}{7} \div \dfrac{15}{26}$

9. $\dfrac{\frac{16}{3}}{\frac{12}{13}}$
                         10. $\dfrac{\frac{14}{2}}{\frac{2}{5}}$

Objective 2 Divide mixed numbers.
For extra help, see Example 3 on pages 309–311 of your text and the Section 5.4 lecture video.
Estimate, then find the actual quotient expressed as a mixed number.

11. $2\dfrac{1}{2} \div 1\dfrac{8}{9}$

12. $3\dfrac{3}{4} \div 2\dfrac{8}{9}$

13. $4\dfrac{1}{4} \div 1\dfrac{4}{7}$

14. $\dfrac{-10\dfrac{1}{4}}{-2\dfrac{3}{5}}$

Objective 3 Divide rational expressions.
For extra help, see Example 4 on page 311 of your text and the Section 5.4 lecture video.
Divide. Write the quotient in lowest terms.

15. $\dfrac{50x^6}{2y^6} \div \dfrac{625x^3}{10y^2}$

16. $\dfrac{28w^{10}}{7x^8 z} \div \dfrac{-16w^8}{14x^5 z}$

Objective 4 Find the square root of a fraction.
For extra help, see Example 5 on page 312 of your text and the Section 5.4 lecture video.
Simplify.

17. $\sqrt{\dfrac{25}{36}}$

18. $\sqrt{\dfrac{25}{49}}$

19. $\sqrt{\dfrac{18}{2}}$

20. $\sqrt{\dfrac{45}{5}}$

Objective 5 Solve equations involving fractions.
For extra help, see Examples 6–7 on pages 312–313 of your text and the Section 5.4 lecture video.
Solve and check.

21. $\dfrac{-5}{7}z = -20$

22. $\dfrac{y}{7} = \dfrac{-3}{2}$

Objective 6 Solve applications involving division of fractions.
For extra help, see Example 8 on pages 314–315 of your text and the Section 5.4 lecture video.
Solve.

23. A cream is sold in a 34-gram container. The average amount of cream used per application is $3\dfrac{2}{5}$ grams. How many applications can be made with the container?

24. A sporting goods manufacturer requires $\dfrac{5}{3}$ yards of fabric to make a pair of soccer shorts. How many shorts can be made from 15 yards of fabric?

Chapter 5 FRACTIONS AND RATIONAL EXPRESSIONS

5.5 Least Common Multiple

Learning Objectives
1   Find the least common multiple (LCM) by listing.
2   Find the LCM using prime factorization.
3   Find the LCM of a set of monomials.
4   Write fractions as equivalent fractions with the least common denominator (LCD).
5   Write rational expressions as equivalent expressions with the LCD.

Key Terms
Use the vocabulary terms listed below to complete each statement in exercises 1-3.

least	**greatest**	**multiple**	**factor**
numerator	**denominator**	**divisor**	**quotient**

1.   The LCD is the least common _____ of the denominators.

2.   The _____ common multiple of a set of numbers is divisible by all the numbers in the set.

3.   To find the LCM using prime factorization, find the prime factorization of each given number then write a factorization that contains each prime factor the _____ number of times it occurs in the factorizations.

GUIDED EXAMPLES AND PRACTICE

Objective 1   Find the least common multiple (LCM) by listing.

Review this example for Objective 1:
1.   Find the least common multiple of 18 and 24 by listing.

24 is not evenly divisible by 18, so go to the next multiple of 24.
48 is not evenly divisible by 18, so go to the next multiple of 24
72 is evenly divisible by 18, so it is the LCM.

Practice this exercise:
1.   Find the least common multiple of 21 and 35 by listing.

Objective 2 Find the LCM using prime factorization.

Review this example for Objective 2:

2. Find the least common multiple of 15, 24 and 28 by prime factorization.

$$15 = 3 \cdot 5$$

$$24 = 2^3 \cdot 3$$

$$28 = 2^2 \cdot 7$$

$$\text{LCM}(15, 24, 28) = 2^3 \cdot 3 \cdot 5 \cdot 7$$

2^3 has the greatest exponent for 2, 3 has the greatest exponent for 3, 5 has the greatest exponent for 5, and 7 has the greatest exponent for 7.

$$= 8 \cdot 3 \cdot 5 \cdot 7$$

$$= 840$$

Practice this exercise:

2. Find the least common multiple of 56, 72, and 80 by prime factorization.

Objective 3 Find the LCM from a set of monomials.

Review this example for Objective 3:

3. Find the least common multiple of $25h^2k$ and $35hj^3k$.

$$25h^2k = 5^2 \cdot h^2 \cdot k$$

$$35hj^3k = 5 \cdot 7 \cdot h \cdot j^3 \cdot k$$

$$\text{LCM}(25h^2k, 35hj^3k) = 5^2 \cdot 7 \cdot h^2 \cdot j^3 \cdot k$$

5^2 has the greatest exponent for 5, 7 has the greatest exponent for 7, h^2 has the greatest exponent for h, j^3 has the greatest exponent for j, and k has the greatest exponent for k.

$$5^2 \cdot 7 \cdot h^2 \cdot j^3 \cdot k = 175h^2 j^3 k$$

Practice this exercise:

3. Find the least common multiple of $24b^2d^2$ and $54b^2c^3d$.

Objective 4 Write fractions as equivalent fractions with the least common denominator (LCD).

Review this example for Objective 4:

4. Write $\dfrac{3}{8}$ and $\dfrac{7}{12}$ as equivalent fractions with the least common denominator.

Practice this exercise:

4. Write $\dfrac{5}{6}$ and $\dfrac{13}{15}$ as equivalent fractions with the least common denominator.

12 is not evenly divisible by 8, so go
to the next multiple of 12.
24 is evenly divisible by 8. Write
equivalent fractions using 24 as the
denominator.

To find the factor that multiplies by 8
to equal 24, divide 24 by 8.

$$24 \div 8 = 3$$

To write the equivalent fraction,
multiply the numerator and
denominator by 3.

$$\frac{3}{8} = \frac{3 \cdot 3}{8 \cdot 3} = \frac{9}{24}$$

To find the factor that multiplies by
12 to equal 24, divide 24 by 12.

$$24 \div 12 = 2$$

To write the equivalent fraction,
multiply the numerator and
denominator by 2.

$$\frac{7}{12} = \frac{7 \cdot 2}{12 \cdot 2} = \frac{14}{24}$$

Objective 5 Write rational expressions as equivalent expressions with the LCD.

Review this example for Objective 5:

5.   Write $\dfrac{4}{9t^3}$ and $\dfrac{5}{12t^2}$ as equivalent
rational expressions with the LCD.

The LCM of 9 and 12 is 36. For the
variables, t^3 has the greater exponent,
so the LCD is $36t^3$. Write equivalent
rational expressions with $36t^3$ as the
denominator.

Practice this exercise:

5.   Write $\dfrac{2}{7m^5}$ and $\dfrac{6}{11m^3}$ as equivalent
rational expressions with the LCD.

To find the factor that multiplies by $9t^3$ to equal $36t^3$, divide $36t^3$ by $9t^3$.

$$36t^3 \div 9t^3 = 4$$

To write the equivalent rational expressions, multiply the numerator and denominator by 4.

$$\frac{4}{9t^3} = \frac{4 \cdot 4}{9t^3 \cdot 4} = \frac{16}{36t^3}$$

To find the factor that multiplies by $12t^2$ to equal $36t^3$, divide $36t^3$ by $12t^2$.

$$36t^3 \div 12t^2 = 3t$$

To write the equivalent rational expressions, multiply the numerator and denominator by $3t$.

$$\frac{5}{12t^2} = \frac{5 \cdot 3t}{12t^2 \cdot 3t} = \frac{15t}{36t^3}$$

ADDITIONAL EXERCISES

Objective 1 Find the least common multiple (LCM) by listing.
For extra help, see Example 1 on pages 318–319 of your text and the Section 5.5 lecture video.
Find the LCM by listing.

1. 12 and 15

2. 21 and 63

3. 9, 15, and 25

4. 4, 14, and 49

Objective 2 Find the LCM using prime factorization.
For extra help, see Example 2 on page 320 of your text and the Section 5.5 lecture video.
Find the LCM using prime factorization.

5. 18 and 12

6. 98 and 28

7. 147 and 63

8. 105 and 189

9. 27, 14, and 63   **10.** 8, 15, and 20

Objective 3 Find the LCM of a set of monomials.
For extra help, use Example 3 on page 320 of your text and the Section 5.5 lecture video.
Find the LCM.

11. $9x$ and $25y$ **12.** $4x$ and $49y$

13. $147x^9z^4$ and $49x^5z^2$ **14.** $28y^9x^6$ and $4y^3x^7$

Objective 4 Write fractions as equivalent fractions with the least common denominator (LCD).
For extra help, see Exercise 4 on page 321 of your text and the Section 5.5 lecture video.
Write as equivalent fractions with the LCD.

15. $\dfrac{11}{21}$ and $\dfrac{7}{18}$ **16.** $\dfrac{5}{6}$ and $\dfrac{3}{4}$

17. $\dfrac{13}{15}$ and $\dfrac{41}{45}$ **18.** $\dfrac{3}{4}$, $\dfrac{5}{14}$ and $\dfrac{43}{49}$

Objective 5 Write rational expressions as equivalent expressions with the LCD.
For extra help, see Exercise 5 on page 322 of your text and the Section 5.5 lecture video.
Write as equivalent expressions with the LCD.

19. $\dfrac{2}{9x}$ and $\dfrac{11}{15y}$ **20.** $\dfrac{178}{245z^9x^2}$ and $\dfrac{31m}{49z^2x^6}$

21. $\dfrac{94}{147x^6y^2}$ and $\dfrac{41}{63x^4y^8}$ **22.** $-\dfrac{43}{98y^9x^4}$ and $\dfrac{41m}{49y^3x^9}$

Chapter 5 FRACTIONS AND RATIONAL EXPRESSIONS

5.6 Adding and Subtracting Fractions, Mixed Numbers, and Rational Expressions

Learning Objectives
1   Add and subtract fractions with the same denominator.
2   Add and subtract rational expressions with the same denominator.
3   Add and subtract fractions with different denominators.
4   Add and subtract rational expressions with different denominators.
5   Add and subtract mixed numbers.
6   Add and subtract negative mixed numbers.
7   Solve equations.
8   Solve applications.

GUIDED EXAMPLES AND PRACTICE

Objective 1 Add and subtract fractions with the same denominator.

Review these examples for Objective 1:

1.   Add $\dfrac{1}{12} + \dfrac{7}{12}$.

$\dfrac{1}{12} + \dfrac{7}{12} = \dfrac{1+7}{12}$ Add the numerators and keep the same denominator.

$= \dfrac{8}{12}$

$= \dfrac{2}{3}$ Simplify.

2.   Subtract $\dfrac{3}{4} - \dfrac{1}{4}$.

$\dfrac{3}{4} - \dfrac{1}{4} = \dfrac{3-1}{4}$ Subtract the numerators and keep the same denominator.

$= \dfrac{2}{4}$ Simplify.

$= \dfrac{1}{2}$

Practice these exercises:

1.   Add $\dfrac{-3}{8} + \dfrac{5}{8}$.

2.   Subtract $\dfrac{8}{9} - \dfrac{4}{9}$.

Objective 2 Add and subtract rational expressions with the same denominator.

Review these examples for Objective 2:

3. Add $\dfrac{5v}{16} + \dfrac{3v}{16}$.

$$\dfrac{5v}{16} + \dfrac{3v}{16} = \dfrac{5v + 3v}{16}$$ Add the numerators and keep the same denominator.

$$= \dfrac{8v}{16}$$

$$= \dfrac{v}{2}$$ Simplify.

4. Subtract $\dfrac{6}{5n} - \dfrac{9}{5n}$.

$$\dfrac{6}{5n} - \dfrac{9}{5n} = \dfrac{6 - 9}{5n}$$ Subtract the numerators and keep the same denominator.

$$= \dfrac{-3}{5n}$$

Practice these exercises:

3. Add $\dfrac{5s}{7} + \dfrac{s}{7}$.

4. Subtract $\dfrac{7b}{10} - \dfrac{3}{10}$.

Objective 3 Add and subtract fractions with different denominators.

Review this example for Objective 3:

5. Add $\dfrac{5}{12} + \dfrac{4}{9}$.

The LCD for 9 and 12 is 36. To write equivalent fractions with 36 as the denominator, rewrite $\dfrac{5}{12}$ by multiplying its numerator and denominator by 3, and rewrite $\dfrac{4}{9}$ by multiplying its numerator and denominator by 4.

Practice this exercise:

5. Subtract $\dfrac{5}{6} - \dfrac{7}{15}$.

$$\frac{5}{12}+\frac{4}{9}=\frac{5(3)}{12(3)}+\frac{4(4)}{9(4)}$$

$$=\frac{15}{36}+\frac{16}{36}$$

$$=\frac{15+16}{36}$$

$$=\frac{31}{36}$$

Objective 4 Add and subtract rational expressions with different denominators.

Review these examples for Objective 4:

6. Add $\dfrac{2k}{5}+\dfrac{3}{7k}$.

The LCD for 5 and $7k$ is $35k$. To write equivalent fractions with $35k$ as the denominator, rewrite $\dfrac{2k}{5}$ by multiplying its numerator and denominator by $7k$, and rewrite $\dfrac{3}{7k}$ by multiplying its numerator and denominator by 5.

$$\frac{2k(7k)}{5(7k)}+\frac{3(5)}{7k(5)}=\frac{14k^2}{35k}+\frac{15}{35k}$$

$$=\frac{14k^2+15}{35k}$$

Because $14k^2$ and 15 are not like terms, express their sum as a polynomial. Because $14k^2+15$ cannot be factored, the rational expression cannot be simplified.

7. Subtract $\dfrac{11}{12c}-\dfrac{13}{18c}$.

The LCD of $12c$ and $18c$ is $36c$.

Practice these exercises:

6. Add $\dfrac{3}{10w}+\dfrac{3}{8}$.

7. Subtract $\dfrac{11z}{20}-\dfrac{7}{30z}$.

To write equivalent fractions with $36c$ as the denominator, rewrite $\dfrac{11}{12c}$ by multiplying its numerator and denominator by 3, and rewrite $\dfrac{13}{18c}$ by multiplying its numerator and denominator by 2.

$$\frac{11(3)}{12c(3)} - \frac{13(2)}{18c(2)} = \frac{33}{36c} - \frac{26}{36c}$$

$$= \frac{33-26}{36c}$$

$$= \frac{7}{36c}$$

Objective 5 Add and subtract mixed numbers.

Review these examples for Objective 5:

8. Estimate $2\dfrac{1}{4} + 5\dfrac{4}{5}$; then find the sum expressed as a mixed number in lowest terms.

$2\dfrac{1}{4}$ rounds down to 2, and $5\dfrac{4}{5}$ rounds up to 6, so the actual sum should be around $2+6=8$.

Notice that the LCD is 20.

$2\dfrac{1}{4} + 5\dfrac{4}{5} = \dfrac{9}{4} + \dfrac{29}{5}$ Write as improper fractions.

$\qquad = \dfrac{9(5)}{4(5)} + \dfrac{29(4)}{5(4)}$

$\qquad = \dfrac{45}{20} + \dfrac{116}{20}$

$\qquad = \dfrac{161}{20}$ Add the numerators.

$\qquad = 8\dfrac{1}{20}$ Write the result as a mixed number.

The actual sum agrees with our estimate.

Practice these exercises:

8. Estimate $4\dfrac{1}{4} + 4\dfrac{2}{3}$; then find the sum expressed as a mixed number in lowest terms.

9. Estimate $4\frac{1}{6}-3\frac{2}{9}$; then find the difference expressed as a mixed number in lowest terms.

$4\frac{1}{6}$ rounds down to 4, and $3\frac{2}{9}$ rounds down to 3, so the actual difference should be around $4-3=1$.

Notice that the LCD is 18.

$$4\frac{1}{6}-3\frac{2}{9}=\frac{25}{6}-\frac{29}{9}$$ Write as improper fractions.

$$=\frac{25(3)}{6(3)}-\frac{29(2)}{9(2)}$$

$$=\frac{75}{18}-\frac{58}{18}$$

$$=\frac{17}{18}$$ Subtract.

The actual difference agrees with our estimate.

9. Estimate $3\frac{5}{6}-\frac{7}{8}$; then find the difference expressed as a mixed number in lowest terms.

Objective 6 Add and subtract negative mixed numbers.

Review this example for Objective 6:

10. Add $3\frac{1}{3}+\left(-4\frac{2}{5}\right)$.

Notice that the LCD is 15.

Practice this exercise:

10. Subtract $-2\frac{7}{8}-\left(-5\frac{1}{4}\right)$.

$$3\frac{1}{3} + \left(-4\frac{2}{5}\right) = \frac{10}{3} + \left(-\frac{22}{5}\right)$$

Write the mixed numbers as improper fractions.

$$= \frac{10(5)}{3(5)} + \left[-\frac{22(3)}{5(3)}\right]$$

$$= \frac{50}{15} + \left(-\frac{66}{15}\right)$$

$$= \frac{50 + (-66)}{15}$$

$$= \frac{-16}{15} \qquad \text{Add.}$$

$$= -1\frac{1}{15} \qquad \begin{array}{l}\text{Write the result as} \\ \text{a mixed number.}\end{array}$$

Objective 7 Solve equations.

Review this example for Objective 7:

11. Solve and check. $r + \frac{1}{5} = \frac{2}{7}$.

Notice that the LCD is 35.

$$r + \frac{1}{5} - \frac{1}{5} = \frac{2}{7} - \frac{1}{5} \qquad \begin{array}{l}\text{Subtract } \frac{1}{5} \text{ from both} \\ \text{sides to isolate } r.\end{array}$$

$$r + 0 = \frac{2(5)}{7(5)} - \frac{1(7)}{5(7)} \qquad \begin{array}{l}\text{Write equivalent} \\ \text{fractions with} \\ \text{the LCD, 35.}\end{array}$$

$$r = \frac{10}{35} - \frac{7}{35}$$

$$r = \frac{3}{35} \qquad \text{Subtract.}$$

Practice this exercise:

11. Solve and check. $f - \frac{5}{8} = \frac{1}{12}$.

Check: $r + \dfrac{1}{5} = \dfrac{2}{7}$

$$\dfrac{3}{35} + \dfrac{1}{5} \overset{?}{=} \dfrac{2}{7}$$ Replace x with $\dfrac{3}{35}$.

$$\dfrac{3}{35} + \dfrac{1(7)}{5(7)} \overset{?}{=} \dfrac{2}{7}$$ Write equivalent fractions with the LCD, 35.

$$\dfrac{3}{35} + \dfrac{7}{35} \overset{?}{=} \dfrac{2}{7}$$

$$\dfrac{10}{35} \overset{?}{=} \dfrac{2}{7}$$ Add the fractions.

$$\dfrac{2}{7} = \dfrac{2}{7}$$ Simplify. The equation is true; so $\dfrac{3}{35}$ is a solution.

Objective 8   Solve applications.

Review these examples for Objective 8:

12. A computer help company is taking job applications, where $\dfrac{5}{9}$ of the applicants have less than 2 years experience doing computer help, and $\dfrac{4}{15}$ of the applicants have more than 5 years experience doing computer help. What fraction of the applicants have between 2 and 5 years of experience?

There are three distinct categories into which an applicant can fall: less than 2 years experience, between 2 and 5 years experience, and more than 5 years experience. Let a represent the number of applicants with between 2 and 5 years of experience.

Practice these exercises:

12. One car gets $29\dfrac{3}{10}$ miles per gallon on the highway. A second car gets $35\dfrac{4}{15}$ miles per gallon on the highway. How many more miles per gallon does the second car get on the highway than the first car?

$$\frac{5}{9}+\frac{4}{15}+a=1$$

$$\frac{5(5)}{9(5)}+\frac{4(3)}{15(3)}+a=\frac{1(45)}{1(45)}$$

Write equivalent fractions with the LCD, 45.

$$\frac{25}{45}+\frac{12}{45}+a=\frac{45}{45}$$

$$\frac{37}{45}+a=\frac{45}{45}$$

Combine fractions on the left side.

$$\frac{37}{45}+a-\frac{37}{45}=\frac{45}{45}-\frac{37}{45}$$

Subtract $\frac{37}{45}$ from both sides to isolate a.

$$a=\frac{8}{45}$$

$\frac{8}{45}$ of the applicants have between 2 and 5 years of experience.

ADDITIONAL EXERCISES
Objective 1 Add and subtract fractions with the same denominator.
For extra help, see Examples 1–3 on pages 324–325 of your text and the Section 5.6 lecture video.
Add or subtract.

1. $\dfrac{3}{5}+\dfrac{1}{5}$

2. $\dfrac{4}{15}+\dfrac{9}{15}$

3. $-\dfrac{3}{14}+\left(-\dfrac{8}{14}\right)$

4. $\dfrac{17}{20}-\dfrac{4}{20}$

Objective 2 Add and subtract rational expressions with the same denominator.
For Extra help, see Examples 4–5 on pages 326–327 of your text and the Section 5.6 lecture video.
Add or subtract.

5. $\dfrac{6}{y}+\dfrac{2}{y}$

6. $\dfrac{7}{mn}+\dfrac{11}{mn}$

7. $\dfrac{3p}{16}+\dfrac{4p}{16}$

8. $\dfrac{8q}{h^2}-\dfrac{3q}{h^2}$

9. $\dfrac{4}{21m}-\dfrac{n}{21m}$

10. $\dfrac{3m}{8k}+\dfrac{5n}{8k}$

Objective 3 Add and subtract fractions with different denominators.
For extra help, see Example 6 on page 327 of your text and the Section 5.6 lecture video.
Add or subtract.

11. $\dfrac{2}{11}+\dfrac{1}{33}$

12. $\dfrac{7}{12}-\dfrac{2}{9}$

13. $\dfrac{1}{12}-\dfrac{5}{24}$

14. $\dfrac{2}{3}+\dfrac{1}{5}+\dfrac{3}{8}$

Objective 4 Add and subtract rational expressions with different denominators.
For extra help, see Example 7 on page 328 of your text and the Section 5.6 lecture video.
Add or subtract.

15. $\dfrac{9}{7x}+\dfrac{5}{14x}$

16. $\dfrac{13}{22}+\dfrac{4}{5q}$

17. $\dfrac{4}{7y}-\dfrac{2}{21}$

18. $\dfrac{3}{4b}-\dfrac{5}{12b^3}$

Objective 5 Add and subtract mixed numbers.
For extra help, see Examples 8–9 on pages 329–331 of your text and the Section 5.6 lecture video.
Add or subtract.

19. $6\dfrac{2}{3}+8$

20. $9\dfrac{5}{7}-4$

21. $7\dfrac{1}{8}-\dfrac{3}{8}$

22. $9\dfrac{5}{11}-5\dfrac{3}{5}$

23. $1\dfrac{1}{9}+5\dfrac{1}{2}$

24. $4\dfrac{1}{2}-7\dfrac{3}{4}$

25. $3\dfrac{4}{5}-8\dfrac{5}{6}$

26. $-4\dfrac{2}{5}-9\dfrac{1}{2}$

Objective 6 Add and subtract negative mixed numbers.
For extra help, see Example 10 on pages 331–332 of your text and the Section 5.6 lecture video.
Add or subtract.

27. $7\dfrac{6}{7} + \left(-4\dfrac{1}{3}\right)$

28. $-3\dfrac{8}{15} + 2\dfrac{4}{5}$

29. $-7\dfrac{4}{9} - 8\dfrac{5}{18}$

30. $-1\dfrac{3}{5} - \left(-4\dfrac{2}{3}\right)$

Objective 7 Solve equations.
For extra help, see Example 11 on page 332 of your text and the Section 5.6 lecture video.
Solve and check.

31. $y + \dfrac{4}{5} = \dfrac{11}{20}$

32. $\dfrac{7}{9} + m = \dfrac{4}{21}$

33. $\dfrac{5}{11} = q - \dfrac{3}{4}$

34. $-\dfrac{9}{13} + p = -\dfrac{3}{26}$

Objective 8 Solve applications.
For extra help, see Example 12 on page 33 of your text and the Section 5.6 lecture video.
Solve.

35. A new long-life tire has a tread depth of $\dfrac{7}{16}$ inch, instead of the more typical $\dfrac{13}{32}$ inch. How much deeper is the new tire tread?

36. A parent died and left an estate to four children. One inherited $\dfrac{1}{9}$ of the estate, the second inherited $\dfrac{15}{81}$, and the third inherited $\dfrac{16}{27}$. How much did the fourth child inherit?

37. A book cover is $6\dfrac{4}{5}$ inches by $8\dfrac{19}{25}$ inches. What is the total distance around (perimeter) the cover of the book?

38. Phil biked $2\dfrac{3}{4}$ miles from his house to the town library. He then biked $3\dfrac{1}{6}$ miles from the library to the local market. His house is $4\dfrac{7}{10}$ miles away from the local market. If Phil had biked straight from his house to the market, how many fewer miles would he have biked?

Name: Date:
Instructor: Section:

Chapter 5 FRACTIONS AND RATIONAL EXPRESSIONS

5.7 Order of Operations; Evaluating and Simplifying Expressions

Learning Objectives
1   Use the order of operations agreement to simplify expressions containing fractions and mixed numbers.
2   Find the mean, median and mode of a set of values.
3   Evaluate expressions.
4   Find the area of a trapezoid.
5   Find the area of a circle.
6   Simplify polynomials containing fractions.

GUIDED EXERCISES AND PRACTICE

Objective 1   Use the order of operations agreement to simplify expressions containing fractions and mixed numbers.

Review these examples for Objective 1:

1.   Simplify $3\frac{1}{5} - 1\frac{2}{3} \cdot \left(-\frac{4}{5}\right)$.

$$3\frac{1}{5} - 1\frac{2}{3} \cdot \left(-\frac{4}{5}\right) = \frac{16}{5} - \frac{5}{3} \cdot \left(-\frac{4}{5}\right)$$

Write the mixed numbers as improper fractions.

$$= \frac{16}{5} - \frac{\overset{1}{\cancel{5}}}{3} \cdot \left(-\frac{4}{\underset{1}{\cancel{5}}}\right)$$

Divide out the common factor of 5.

$$= \frac{16}{5} - \left(-\frac{4}{3}\right)$$

$$= \frac{16(3)}{5(3)} - \left(-\frac{4(5)}{3(5)}\right)$$

Write equivalent fractions with the LCD, 15.

$$= \frac{48}{15} - \left(-\frac{20}{15}\right)$$

$$= \frac{68}{15} \qquad \text{Subtract.}$$

$$= 4\frac{8}{15} \qquad \text{Write as a mixed number.}$$

Practice these exercises:

1.   Simplify $-\frac{5}{12} + 1\frac{3}{4} \cdot \frac{2}{3}$.

2. Simplify $3\left(\dfrac{1}{6}+\dfrac{1}{9}\right)-\left(\dfrac{3}{4}\right)^2$.

$$3\left(\frac{1}{6}+\frac{1}{9}\right)-\left(\frac{3}{4}\right)^2 = 3\left(\frac{1(3)}{6(3)}+\frac{1(2)}{9(2)}\right)-\left(\frac{3}{4}\right)^2$$

Write equivalent fractions with the LCD, 18.

$$= 3\left(\frac{3}{18}+\frac{2}{18}\right)-\left(\frac{3}{4}\right)^2$$

$$= 3\cdot\frac{5}{18}-\frac{9}{16}$$

Add and evalute the expression with the

exponent.

$$= \frac{\overset{1}{\cancel{3}}}{1}\cdot\frac{5}{\underset{1}{\cancel{3}\cdot 6}}-\frac{9}{16}$$

Divide out the common factors.

$$= \frac{5}{6}-\frac{9}{16}\qquad\text{Multiply.}$$

$$= \frac{5(8)}{6(8)}-\frac{9(3)}{16(3)}$$

Write equivalent fractions with the LCD, 48.

$$= \frac{40}{48}-\frac{27}{48}$$

$$= \frac{13}{48}\qquad\text{Subtract.}$$

2. Simplify $\sqrt{\dfrac{25}{4}}-3\left(\dfrac{2}{5}+\dfrac{1}{2}\right)$.

Objective 2   Find the mean, median and mode of a set of values.

Review this example for Objective 2:

3. A practicing shot putter puts her shot $10\dfrac{1}{3}$ m, $15\dfrac{5}{6}$ m, $12\dfrac{1}{12}$ m, $11\dfrac{1}{3}$ m, and $15\dfrac{5}{6}$ m. Find the mean, median, and mode distance.

Practice this exercise:

3. A teacher times his students taking a short test. Seven students took $7\dfrac{1}{3}$ minutes, $5\dfrac{1}{3}$ minutes, $4\dfrac{5}{6}$ minutes, $6\dfrac{1}{6}$ minutes, $7\dfrac{2}{3}$ minutes,

$$\text{mean} = \left(10\frac{1}{3} + 15\frac{5}{6} + 12\frac{1}{12} + 11\frac{1}{3} + 15\frac{5}{6}\right) \div 5$$

Divide the sum of the distances by the number of distances, 5.

$$= \left(10\frac{4}{12} + 15\frac{10}{12} + 12\frac{1}{12} + 11\frac{4}{12} + 15\frac{10}{12}\right) \div 5$$

Write each fraction with the LCD, 12.

$$= 63\frac{29}{12} \div 5 \qquad \text{Add the mixed numbers.}$$

$$= \frac{785}{12} \cdot \frac{1}{5}$$

Write the mixed number as an improper fraction and write the division as multiplication.

$$= \frac{157 \cdot \cancel{5}^{1}}{12} \cdot \frac{1}{\cancel{5}_{1}} \qquad \begin{array}{l} \text{Divide out the common} \\ \text{factor, 5.} \end{array}$$

$$= \frac{157}{12}, \text{ or } 13\frac{1}{12} \qquad \text{Multiply.}$$

The mean distance is $13\frac{1}{2}$ m.

Median and mode:

$$10\frac{1}{3}, 11\frac{1}{3}, 12\frac{1}{12}, 15\frac{5}{6}, 15\frac{5}{6}$$

This list has an odd number of values, so the middle value, $12\frac{1}{12}$ m, is the median.
The mode is the number that occurs most often, which is $15\frac{5}{6}$ m.

minutes, $5\frac{1}{3}$ minutes, and

$6\frac{1}{2}$ minutes to finish the test.

a. Find the mean test time of these students.
b. Find the median test time of these students.
c. Find the mode test time of these students.

Objective 3   Evaluate expressions.

Review this example for Objective 3:

4.   Evaluate $7x - yz$ when $x = \dfrac{3}{4}$, $y = 1\dfrac{1}{3}$, and

$z = 2\dfrac{1}{5}$.

$7\left(\dfrac{3}{4}\right) - \left(1\dfrac{1}{3}\right)\left(2\dfrac{1}{5}\right) = \dfrac{7}{1}\left(\dfrac{3}{4}\right) - \left(\dfrac{4}{3}\right)\left(\dfrac{11}{5}\right)$

Replace x with $\dfrac{3}{4}$, y with $1\dfrac{1}{3}$, and z with $2\dfrac{1}{5}$,

then rewrite mixed numbers as improper

fractions.

$= \dfrac{21}{4} - \dfrac{44}{15}$ Multiply.

$= \dfrac{21(15)}{4(15)} - \dfrac{44(4)}{15(4)}$

Write equivalent fractions with the LCD, 60.

$= \dfrac{315}{60} - \dfrac{176}{60}$

$= \dfrac{139}{60}$, or $2\dfrac{19}{60}$ Subtract.

Practice this exercise:

4.   Evaluate $b^2 - 4ac$ when

$a = 1\dfrac{1}{3}$, $b = 2\dfrac{2}{3}$, and

$c = \dfrac{5}{8}$.

Objective 4   Find the area of a trapezoid.

Review this example for Objective 4:

5.   Find the area of a trapezoid with height $3\dfrac{2}{5}$ cm,

smaller base length 8 cm, and larger base length

$9\dfrac{1}{2}$ cm.

Use the formula $A = \dfrac{1}{2}h(a+b)$.

Practice this exercise:

5.   Find the area of a

trapezoid with height $5\dfrac{1}{4}$

ft, smaller base length 12

ft, and larger base length

15 ft.

$$A = \frac{1}{2}\left(3\frac{2}{5}\right)\left(8 + 9\frac{1}{2}\right)$$

Replace h with $3\frac{2}{5}$, a with 8, and b with $9\frac{1}{2}$.

$$A = \frac{1}{2}\left(3\frac{2}{5}\right)\left(17\frac{1}{2}\right)$$

Add inside the parentheses.

$$A = \frac{1}{2}\left(\frac{17}{5}\right)\left(\frac{35}{2}\right)$$

Write as improper fractions.

$$A = \frac{1}{2}\left(\frac{17}{\cancel{5}_1}\right)\left(\frac{\cancel{5}^1 \cdot 7}{2}\right)$$

Divide out the common factor.

$$A = \frac{119}{4}$$

Multiply.

$$A = 29\frac{3}{4}$$

The area is $29\frac{3}{4}$ cm^2.

Objective 5   Find the area of a circle.

Review this example for Objective 5:

6. Find the area of a circle with diameter 13 cm. Use $\frac{22}{7}$ for π.

$$r = \frac{1}{2}d$$

$$r = \frac{1}{2}(13) \qquad \text{Replace } d \text{ with } 13.$$

$$r = \frac{1}{2} \cdot \frac{13}{1}$$

$$r = \frac{13}{2}$$

Practice this exercise:

6. Find the area of a circle with radius $\frac{3}{4}$ m. Use $\frac{22}{7}$ for π.

$$A = \pi r^2$$

$$A \approx \frac{22}{7} \cdot \left(\frac{13}{2}\right)^2 \qquad \text{Replace } \pi \text{ with } \frac{22}{7} \text{ and } r$$
$$\text{with } \frac{13}{2}.$$

$$A \approx \frac{22}{7} \cdot \frac{169}{4} \qquad \text{Square } \frac{13}{2} \text{ to get } \frac{169}{4}.$$

$$A \approx \frac{\overset{11}{\cancel{22}}}{7} \cdot \frac{169}{\underset{2}{\cancel{4}}} \qquad \text{Divide out the common}$$
$$\text{factor of 2.}$$

$$A \approx \frac{1859}{14} \qquad \text{Multiply.}$$

$$A \approx 132\frac{11}{14} \qquad \text{Write the result as a mixed}$$
$$\text{number.}$$

The area is $132\frac{11}{14}$ cm^2.

Objective 6 Simplify polynomials containing fractions.

Review these examples for Objective 6:

7. Combine like terms. $\dfrac{4}{9}w + \dfrac{11}{12}w^2 - \dfrac{1}{8}w^2 + \dfrac{1}{6}w$

$$\frac{4}{9}w + \frac{11}{12}w^2 - \frac{1}{8}w^2 + \frac{1}{6}w$$

$$= \frac{11}{12}w^2 - \frac{1}{8}w^2 + \frac{4}{9}w + \frac{1}{6}w$$

Use the commutative property to group the like terms together. Note that this step is optional.

$$= \frac{11(2)}{12(2)}w^2 - \frac{1(3)}{8(3)}w^2 + \frac{4(2)}{9(2)}w + \frac{1(3)}{6(3)}w$$

Write equivalent fractions. The LCD for the w^2 coefficients is 24. The LCD for the w coefficients is 18.

$$= \frac{22}{24}w^2 - \frac{3}{24}w^2 + \frac{8}{18}w + \frac{3}{18}w$$

$$= \frac{19}{24}w^2 + \frac{11}{18}w \qquad \text{Add or subtract coefficients.}$$

Practice these exercises:

7. Combine like terms.

$$\frac{1}{2}j^3 + \frac{3}{4} - \frac{2}{5}j^2 + \frac{3}{7}j^3 - \frac{3}{10}j^2$$

8. Add or subtract. $\left(\dfrac{7}{10}g - \dfrac{13}{18}\right) - \left(\dfrac{2}{15}g - \dfrac{11}{24}\right)$

$\left(\dfrac{7}{10}g - \dfrac{13}{18}\right) - \left(\dfrac{2}{15}g - \dfrac{11}{24}\right)$

$= \left(\dfrac{7}{10}g - \dfrac{13}{18}\right) + \left(-\dfrac{2}{15}g + \dfrac{11}{24}\right)$

Write an equivalent addition expression.

$= \dfrac{7}{10}g - \dfrac{2}{15}g - \dfrac{13}{18} + \dfrac{11}{24}$

Collect like terms. Again, this is optional.

$= \dfrac{7(3)}{10(3)}g - \dfrac{2(2)}{15(2)}g - \dfrac{13(4)}{18(4)} + \dfrac{11(3)}{24(3)}$

Write equivalent fractions with the LCD.

$= \dfrac{21}{30}g - \dfrac{4}{30}g - \dfrac{52}{72} + \dfrac{33}{72}$

$= \dfrac{17}{30}g - \dfrac{19}{72}$ Combine the like terms.

9. Multiply. $\dfrac{4}{7}\left(\dfrac{21}{22}b + 35\right)$

Use the distributive property.
$\dfrac{4}{7}\left(\dfrac{21}{22}b + 35\right)$

$= \dfrac{4}{7} \cdot \dfrac{21}{22}b + \dfrac{4}{7} \cdot \dfrac{35}{1}$ Distribute the $\dfrac{4}{7}$.

$= \dfrac{2 \cdot \overset{2}{\cancel{2}}}{\cancel{7}} \cdot \dfrac{\overset{1}{\cancel{7}} \cdot 3}{\underset{1}{\cancel{2}} \cdot 11}b + \dfrac{4}{\cancel{7}} \cdot \dfrac{\overset{1}{\cancel{7}} \cdot 5}{1}$ Divide out the common factors.

$= \dfrac{6}{11}b + 20$ Multiply.

10. Multiply. $\left(\dfrac{3}{5}z + 6\right)\left(\dfrac{2}{3}z - 3\right)$

Multiply each term in the second polynomial by each term in the first polynomial.

8. Add or subtract.
$\left(\dfrac{13}{20}u^2 + \dfrac{2}{9}v\right) - \left(-\dfrac{5}{21}v + \dfrac{3}{8}u^2\right)$

9. Multiply.
$\left(-\dfrac{3}{22}n^3\right)\left(-\dfrac{11}{36}n^2\right)$

10. Multiply.
$\left(\dfrac{5}{6}s + 8\right)\left(\dfrac{1}{2}s - 3\right)$

$$\left(\frac{3}{5}z+6\right)\left(\frac{2}{3}z-3\right)$$

$$=\frac{3}{5}z\cdot\frac{2}{3}z+\frac{3}{5}z\cdot(-3)+6\cdot\frac{2}{3}z+6\cdot(-3)$$

$$=\frac{\overset{1}{\cancel{3}}}{5}z\cdot\frac{2}{\cancel{3}}z+\frac{3}{5}z\cdot\frac{-3}{1}+\frac{\overset{2}{\cancel{6}}}{1}\cdot\frac{2}{\cancel{3}}z-18$$

Divide out the common factors.

$$=\frac{2}{5}z^2-\frac{9}{5}z+4z-18 \qquad\qquad \text{Multiply.}$$

$$=\frac{2}{5}z^2-\frac{9}{5}z+\frac{4(5)}{1(5)}z-18$$

Write the coefficients of the like terms with

their LCD, 5.

$$=\frac{2}{5}z^2-\frac{9}{5}z+\frac{20}{5}z-18$$

$$=\frac{2}{5}z^2+\frac{11}{5}z-18$$

Subtract coefficients of the like terms.

ADDITIONAL EXERCISES
Objective 1   Use the order of operations agreement to simplify expressions containing
   fractions and mixed numbers.

For extra help, see Examples 1–2 on pages 337–338 of your text and the Section 5.7
lecture video.

Simplify.

1. $\dfrac{3}{5}\div\left(-\dfrac{6}{5}\right)-3\dfrac{3}{5}$

2. $\dfrac{4}{5}\div\left(-\dfrac{4}{3}\right)-1\dfrac{3}{4}$

3. $2\dfrac{1}{2}-\left(\dfrac{3}{5}\right)^2$

4. $4\dfrac{1}{2}-3\sqrt{\dfrac{50}{2}}$

5. $3\dfrac{1}{4}-3\sqrt{\dfrac{100}{4}}$

6. $\left(\dfrac{2}{3}\right)^2\div\dfrac{1}{5}\left(-4\dfrac{3}{4}+\dfrac{3}{7}\right)$

Objective 2 Find the mean, median and mode of a set of values.
For extra help, see Example 3 on pages 338–339 of your text and the Section 5.7 lecture video.

7. A first grade teacher asks her students about their height and weight. The table lists the information she collected from six of her students.
   a.   Find the mean height and weight.
   b.   Find the median height and weight.
   c.   Find the mode height and weight.

Height (inches)	Weight (pounds)
$46\frac{1}{4}$	$42\frac{1}{4}$
48	$60\frac{1}{2}$
$44\frac{3}{4}$	$42\frac{1}{4}$
$47\frac{1}{2}$	$52\frac{1}{2}$
$45\frac{3}{4}$	$48\frac{1}{2}$
$47\frac{1}{2}$	$51\frac{1}{4}$

8. During a particular half-hour, a radio station plays eight songs and eight ads. The table lists the duration, in minutes, of each song and ad played.
   a.   Find the mean duration of the songs and ads.
   b.   Find the median duration of the songs and ads.
   c.   Find the mode duration of the songs and ads.

Songs	Ads
$2\frac{1}{2}$	$\frac{1}{6}$
$3\frac{1}{6}$	$1\frac{1}{4}$
$2\frac{5}{8}$	$\frac{1}{2}$
$3\frac{1}{10}$	$\frac{3}{4}$
$2\frac{8}{9}$	$\frac{5}{6}$
$2\frac{1}{4}$	$\frac{1}{2}$
$3\frac{3}{8}$	$\frac{1}{3}$
$3\frac{1}{4}$	$1\frac{1}{4}$

Name: Date:
Instructor: Section:

Objective 3 Evaluate expressions.

For extra help, see Example 4 on page 339 of your text and the Section 5.7 lecture video.

Evaluate each expression using the given values.

9. $r + ps; \ r = 2\frac{1}{2}, \ p = 3, \ s = 1\frac{1}{4}$

10. $hr^2; \ h = 3\frac{3}{5}, \ r = \frac{2}{9}$

11. $\frac{1}{2}ap^2; \ a = -10\frac{5}{6}, \ p = \frac{1}{2}$

12. $3lmn; \ l = \frac{4}{7}, \ m = -3\frac{1}{2}, \ n = -\frac{7}{11}$

Objective 4 Find the area of a trapezoid.

For extra help, see Example 5 on page 341 of your text and the Section 5.7 lecture video.

Find the area of the trapezoids shown.

13.

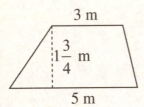

14.

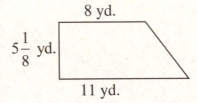

Objective 5 Find the area of a circle.

For extra help, see Example 6 on pages 341–342 of your text and the Section 5.7 lecture video.

Find each area. Use $\frac{22}{7}$ for π.

15. Find the area of a circle with radius of 9 centimeters.

16. Find the area of a circle with a diameter of $5\frac{1}{4}$ feet.

Objective 6 Simplify polynomials containing fractions.

For extra help, see Examples 7–11 on pages 342–344 of your text and the Section 5.7 lecture video.

Combine like terms and write the resulting polynomial in descending order of degree.

17. $\frac{5}{8}y - \frac{1}{6}y^3 + 7 - \frac{3}{4}y^3 - \frac{1}{4}$

18. $\frac{2}{3}n^3 - \frac{3}{5}n + \frac{5}{8}n^2 - \frac{1}{4}n - \frac{1}{2} + \frac{1}{6}n^3$

Add or subtract and write the resulting polynomial in descending order of degree.

19. $\left(\frac{1}{4}n^2 - n + \frac{5}{6}\right) + \left(\frac{2}{3}n^2 + \frac{1}{5}n + \frac{3}{4}\right)$

20. $\left(\frac{1}{7}y^4 - y^3 + \frac{2}{5}y - 1\right) - \left(\frac{5}{7}y^4 + \frac{3}{4}y^3 - 3y - \frac{3}{8}\right)$

Multiply.

21. $\left(\dfrac{5}{8}m^3\right)\left(-\dfrac{4}{7}m^5\right)$

22. $\left(\dfrac{4}{5}x^2y\right)\left(\dfrac{15}{16}x^6\right)$

23. $\dfrac{5}{6}\left(3n-\dfrac{4}{7}\right)$

24. $\dfrac{5}{8}y\left(\dfrac{4}{6}y^2+2\right)$

25. $-\dfrac{3}{8}x^3\left(\dfrac{1}{12}x^2-6x+\dfrac{1}{9}\right)$

26. $\dfrac{2}{7}m^2n\left(\dfrac{3}{4}m^3n-\dfrac{7}{8}n^2+3m\right)$

27. $\left(\dfrac{2}{3}x-4\right)\left(3x+\dfrac{3}{2}\right)$

28. $\left(\dfrac{1}{3}y+\dfrac{4}{5}\right)\left(\dfrac{1}{3}y-\dfrac{4}{5}\right)$

Chapter 5 FRACTIONS AND RATIONAL EXPRESSIONS

5.8 Solving Equations

Learning Objectives
1.   Use the LCD to eliminate fractions from equations.
2.   Translate sentences to equations; then solve.
3.   Solve applications involving one unknown.
4.   Solve applications involving two unknowns.

GUIDED EXAMPLES AND PRACTICE

Objective 1   Use the LCD to eliminate fractions from equations.

Review this example for Objective 1:

1.   Solve and check. $\dfrac{3}{5}(r-5) = \dfrac{2}{7}r - \dfrac{1}{2}$

$$\frac{3}{5}(r-5) = \frac{2}{7}r - \frac{1}{2}$$

$$\frac{3}{5}r - \frac{3}{\cancel{5}} \cdot \frac{\cancel{5}^{1}}{1} = \frac{2}{7}r - \frac{1}{2}$$

Distribute $\dfrac{3}{5}$ to eliminate the parentheses. Then divide out common factors.

$$\frac{3}{5}r - 3 = \frac{2}{7}r - \frac{1}{2} \quad \text{Multiply.}$$

$$70\left(\frac{3}{5}r - 3\right) = \left(\frac{2}{7}r - \frac{1}{2}\right)70$$

Eliminate both fractions by multiplying both sides by the LCD, 70.

$$\frac{14 \cdot \cancel{5}^{1}}{1} \cdot \frac{3}{\cancel{5}_{1}}r - 70 \cdot 3 = \frac{10 \cdot \cancel{7}^{1}}{1} \cdot \frac{2}{\cancel{7}_{1}}r - \frac{35 \cdot \cancel{2}^{1}}{1} \cdot \frac{1}{\cancel{2}_{1}}$$

Distribute 70 and divide out the common factors.

$$42r - 210 = 20r - 35 \quad \text{Multiply.}$$

$$\underline{-20r \qquad\qquad -20r} \qquad \text{Subtract } 20r \text{ from both sides.}$$

$$22r - 210 = 0 - 35$$

Practice this exercise:

1.   Solve and check.

$$\frac{4}{7}k + \frac{3}{10} = \frac{4}{5}$$

Name:
Instructor:

Date:
Section:

$$22r - 210 = -35$$

$$\underline{+210 \quad +210}$$ Add 210 to both sides to isolate the $22r$ term.

$$22r - 0 = 175$$

$$\frac{22r}{22} = \frac{175}{22}$$ Divide both sides by 22 to isolate r.

$$1r = \frac{175}{22}$$ Simplify.

$$r = 7\frac{21}{22}$$ Write the solution as a mixed number.

Check:

$$\frac{3}{5}(r-5) = \frac{2}{7}r - \frac{1}{2}$$

$$\frac{3}{5}\left(7\frac{21}{22}-5\right) \overset{?}{=} \frac{2}{7}\left(7\frac{21}{22}\right) - \frac{1}{2}$$

In the original equation, replace r with $7\frac{21}{22}$.

$$\frac{3}{5}\left(\frac{175}{22}-5\right) \overset{?}{=} \frac{2}{7} \cdot \frac{175}{22} - \frac{1}{2}$$

Write $7\frac{21}{22}$ as an improper fraction.

$$\frac{3}{5}\left(\frac{175}{22}-\frac{5(22)}{1(22)}\right) \overset{?}{=} \frac{2}{7} \cdot \frac{175}{22} - \frac{1}{2}$$

Write equivalent fractions using the LCD, 22.

$$\frac{3}{5}\left(\frac{175}{22}-\frac{110}{22}\right) \overset{?}{=} \frac{2}{7} \cdot \frac{175}{22} - \frac{1}{2}$$

$$\frac{3}{5} \cdot \frac{65}{22} \overset{?}{=} \frac{2}{7} \cdot \frac{175}{22} - \frac{1}{2}$$ Subtract.

$$\frac{3}{\cancel{5}} \cdot \frac{\cancel{5} \cdot 13}{22} \overset{?}{=} \frac{\cancel{2}}{\cancel{7}} \cdot \frac{\cancel{7} \cdot 25}{\cancel{2} \cdot 11} - \frac{1}{2}$$ Divide out common factors.

$$\frac{39}{22} \overset{?}{=} \frac{25}{11} - \frac{1}{2}$$ Multiply.

$$\frac{39}{22}\overset{?}{=}\frac{25(2)}{11(2)}-\frac{1(11)}{2(11)}$$

Write equivalent fractions with the LCD, 22.

$$\frac{39}{22}\overset{?}{=}\frac{50}{22}-\frac{11}{22}$$

Subtract. True;

$$\frac{39}{22}=\frac{39}{22}$$

so $7\dfrac{21}{22}$ is the

solution.

Objective 2 Translate sentences to equations; then solve.

Review this example for Objective 2:

2. Two-ninths less than the product of $1\dfrac{2}{5}$ and b is

1 less than $2\dfrac{1}{2}$ of b. Translate to an equation;

then solve for b.

Translate:

Two-ninths less than the product of $1\dfrac{2}{5}$ and b

$$1\frac{2}{5}b-\frac{2}{9}$$

is 1 less than $2\dfrac{1}{2}$ of b.

$$=2\frac{1}{2}b-1$$

$$1\frac{2}{5}b-\frac{2}{9}=2\frac{1}{2}b-1$$

$$90\left(\frac{7}{5}b-\frac{2}{9}\right)=\left(\frac{5}{2}b-1\right)90$$

$$\frac{\overset{18}{\cancel{90}}}{1}\cdot\frac{7}{\underset{1}{\cancel{5}}}b-\frac{\overset{10}{\cancel{90}}}{1}\cdot\frac{2}{\underset{1}{\cancel{9}}}=\frac{\overset{45}{\cancel{90}}}{1}\cdot\frac{5}{\underset{1}{\cancel{2}}}b-90\cdot1$$

$$126b-20=225b-90$$

Practice this exercise:

2. The sum of $3\dfrac{1}{3}$ of n and

$\dfrac{3}{8}$ is the same as $1\dfrac{1}{6}$ less

than $\dfrac{7}{9}$ of n. Translate to

an equation; then solve

for n.

$$225b - 90 = 126b - 20$$

$$\underline{-126b \qquad\quad -126b}$$

$$99b - 90 = 0 - 20$$

$$\underline{+90 \qquad +90}$$

$$99b + 0 = 70$$

$$\frac{99b}{99} = \frac{70}{99}$$

$$b = \frac{70}{99}$$

Objective 3 Solve applications involving one unknown.

Review these examples for Objective 3:

3. Find the height h of a triangle with a base

of $2\frac{1}{4}$ cm and an area of $3\frac{3}{4}$ cm^2.

The given area is for a triangle; so use $A = \frac{1}{2}bh$.

$$3\frac{3}{4} = \frac{1}{2}\left(2\frac{1}{4}\right)h$$

In $A = \frac{1}{2}bh$, replace A with $3\frac{3}{4}$ and

b with $2\frac{1}{4}$.

$$\frac{15}{4} = \frac{1}{2}\cdot\frac{9}{4}h \qquad \begin{array}{l}\text{Write the mixed numbers}\\ \text{as improper fractions.}\end{array}$$

$$\frac{15}{4} = \frac{9}{8}h \qquad \text{Multiply.}$$

$$\overset{5}{\underset{1}{\cancel{15}}}\cdot\overset{2}{\underset{3}{\cancel{8}}}\cdot\frac{\overset{1}{\cancel{8}}}{\underset{1}{\cancel{9}}} = \frac{\overset{1}{\cancel{8}}}{\underset{1}{\cancel{9}}}\cdot\frac{\overset{1}{\cancel{9}}}{\underset{1}{\cancel{8}}}h \qquad \begin{array}{l}\text{Multiply both sides}\\ \text{by } \frac{8}{9} \text{ and divide out}\\ \text{the common factors.}\end{array}$$

$$\frac{10}{3} = 1h \qquad \text{Multiply.}$$

$$3\frac{1}{3} = h \qquad \begin{array}{l}\text{Write the improper fraction}\\ \text{as a mixed number.}\end{array}$$

The height of the triangle is $3\frac{1}{3}$ cm.

Practice these exercises:

3. Find the length a of a trapezoid with a second

length b of $4\frac{3}{8}$ ft, a height

h of 5 ft, and an area A of

$16\frac{9}{16}$ ft^2.

4. For Janet to win a bowling tournament, she must have an average (mean) score of 248 after all five games. In her first four games, she has bowled a 233, a 260, a 257, and a 229. What is the minimum score Janet needs in her final game to win the tournament?

To calculate the mean, divide the sum of the scores by the number of scores. Let g represent the unknown score.

$$\frac{233 + 260 + 257 + 229 + g}{5} = 248$$

$$\frac{979 + g}{5} = 248$$

Add in the numerator.

$$\not{5} \cdot \frac{979 + g}{\not{5}} = 248 \cdot 5$$

Eliminate the denominator by multiplying both sides by 5.

$$979 + g = 1240$$
$$\underline{-979 \qquad -979}$$
$$0 + g = 261$$

Subtract 979 from both sides.

$$g = 261$$

If Janet bowls a 261 in her final game, the mean of her five games will be exactly 248 and she will win the tournament.

4. A new study suggests that students who do an average of at least 35 algebra problems per weekday every week see a full letter grade improvement in their algebra class over those students that do not. If you have done 41, 20, 39, and 37 problems in the first four weekdays of this week, what is the minimum number of problems you should do on the final weekday to keep pace for the grade improvement?

Objective 4 Solve applications involving two unknowns.

Review these examples for Objective 4:

5. Bob and Stacy are weeding their garden. Bob pulls up weeds at $\frac{3}{4}$ the rate Stacy pulls up weeds. If they pull up 84 total weeds, how many weeds did each of them pull up?

Relationship 1: Bob pulls up weeds at $\frac{3}{4}$ the rate

Practice these exercises:

5. The handle of a fishing rod is $\frac{1}{7}$ the length of the pole. If the rod is $5\frac{3}{5}$ feet long, what are the lengths of the handle and the pole?

Stacy pulls up weeds. Let s be the number of weeds Stacy pulled up.

Bob's weeds $= \dfrac{3}{4} s$

Relationship 2: They pull up 84 total weeds.

$$\dfrac{3}{4} s + s = 84$$

$$4\left(\dfrac{3}{4} s + s \right) = (84)4 \qquad \text{Multiply both sides by}$$
$$\qquad\qquad\qquad\qquad\quad 4 \text{ to clear the fraction.}$$

$$4 \cdot \dfrac{3}{4} s + 4 \cdot s = 84 \cdot 4 \qquad \text{Distribute the 4.}$$

$$3s + 4s = 336 \qquad \text{Multiply.}$$

$$7s = 336 \qquad \text{Combine like terms.}$$

$$\dfrac{7s}{7} = \dfrac{336}{7} \qquad \text{Divide both sides by 3}$$
$$\qquad\qquad\qquad \text{to clear the coefficient.}$$

$$s = 48$$

Stacy pulled up 48 weeds. Since Bob pulled up weeds at $\dfrac{3}{4}$ the rate Stacy pulled up weeds, he only pulled up 36 weeds.

6. The length of a pool is $2\dfrac{1}{2}$ times its width. The perimeter of the pool is $71\dfrac{3}{4}$ feet. What are the dimensions of the pool?

Relationship 1: The length of the pool is $2\dfrac{1}{2}$ times its width. Let w be the width of the pool.

Length $= 2\dfrac{1}{2} w$

Relationship 2: The perimeter of the pool is $71\dfrac{3}{4}$ feet.

6. The height of a 54 cm² trapezoid is 10 cm. The larger length of the trapezoid is $2\dfrac{1}{5}$ times the smaller length. What are the two lengths?

$$2w + 2\left(2\frac{1}{2}w\right) = 71\frac{3}{4}$$

$$2w + 2\left(\frac{5}{2}w\right) = \frac{287}{4}$$ Change to improper fractions.

$$2w + \frac{\overset{1}{\cancel{2}}}{1} \cdot \frac{5}{\underset{1}{\cancel{2}}}w = \frac{287}{4}$$ Divide out common factors.

$$7w = \frac{287}{4}$$ Combine like terms.

$$\frac{\overset{1}{\cancel{1}}}{\underset{1}{\cancel{7}}} \cdot \frac{\overset{1}{\cancel{7}}}{\underset{1}{\cancel{1}}}w = \frac{\overset{41}{\cancel{287}}}{4} \cdot \frac{1}{\underset{1}{\cancel{7}}}$$

Multiply both sides by $\frac{1}{7}$ to isolate w, then divide out the common factors.

$$1w = \frac{41}{4}$$ Multiply.

$$w = 10\frac{1}{4}$$

Change the improper fraction to a mixed number.

The width of the pool is $10\frac{1}{4}$ feet. Because length $= 2\frac{1}{2}w$, replace $10\frac{1}{4}$ in for w to find the length of the pool.

$$\text{Length} = 2\frac{1}{2}\left(10\frac{1}{4}\right)$$

$$= \frac{5}{2} \cdot \frac{41}{4}$$ Change to improper fractions.

$$= \frac{205}{8}$$ Multiply.

$$= 25\frac{5}{8}$$ Change to a mixed number.

The length of the pool is $25\frac{5}{8}$ feet.

ADDITIONAL EXERCISES
Objective 1 Use the LCD to eliminate fractions from equations.
For extra help, see Examples 1–2 on pages 348–351 of your text and the Section 5.8 lecture video.
Use the LCD to simplify the equation, then solve and check.

1. $t + \dfrac{2}{7} = \dfrac{5}{14}$

2. $x - \dfrac{1}{6} = -\dfrac{2}{3}$

3. $-\dfrac{4}{7}x = \dfrac{12}{35}$

4. $\dfrac{4}{7}x = \dfrac{8}{35}$

5. $\dfrac{1}{10} + n = \dfrac{5}{6}n - \dfrac{4}{5}$

6. $\dfrac{1}{9} + x = \dfrac{2}{3}x - \dfrac{1}{2}$

7. $\dfrac{3}{4}(y - 2) = \dfrac{1}{6}y - \dfrac{3}{8}$

8. $\dfrac{7}{12}(f - 24) = \dfrac{11}{30}f - \dfrac{5}{18}$

Objective 2 Translate sentences to equations; then solve.
For extra help, see Example 3 on pages 351–352 of your text and the Section 5.8 lecture video.
Translate to an equation, then solve.

9. $\dfrac{1}{3}$ of a number is $25\dfrac{2}{3}$.

10. The product of $5\dfrac{1}{4}$ and a number is $1\dfrac{1}{2}$.

11. $2\dfrac{7}{10}$ less than h is $-4\dfrac{3}{5}$.

12. $6\dfrac{1}{5}$ more than twice k is $-\dfrac{1}{3}$.

13. $\dfrac{1}{6}$ of the sum of a and 3 is equal to $1\dfrac{1}{9}$ added to a.

14. The sum of $\dfrac{2}{5}$ of z and 3 is half the difference of $5z$ and 3.

Objective 3 Solve applications involving one unknown.
For extra help, see Example 4 on pages 352–353 of your text and the Section 5.8 lecture video.
Solve.

15. Find the height of the shape if the area is $7\frac{7}{8}$ square feet.

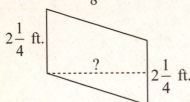

$2\frac{1}{4}$ ft.

?

$2\frac{1}{4}$ ft.

16. Warren's grade in a course is determined by the average (mean) of five tests. His scores on the first four tests are 79, 82, 84, and 80. To get a B he must have a final average of 85 or higher. What is the minimum score he needs on the fifth test to receive a B?

Objective 4 Solve applications involving two unknowns.
For extra help, see Examples 5–6 on pages 353–355 of your text and the Section 5.8 lecture video.
Solve.

17. Two drink coolers hold the same amount of liquid. Cooler A is $\frac{5}{8}$ full while cooler B is $\frac{1}{3}$ full. The two coolers currently have a combined total of 23 fluid ounces of liquid. How much does each cooler hold?

18. In an email inbox, $\frac{1}{4}$ of the messages are marked unread. Of these, $\frac{3}{4}$ are marked junk mail and the rest are not junk. What fraction of the unread messages in the inbox are not junk?

Chapter 6 DECIMALS

6.1 Introduction to Decimal Numbers

Learning Objectives
1   Write decimals as fractions or mixed numbers.
2   Write a word name for a decimal number.
3   Graph decimals on a number line.
4   Use < or > to write a true statement.
5   Round decimal numbers to a specified place.

Key Terms
Use the terms listed below to complete each statement in Exercises 1–4.

| right | fractions | hundredths | left | 4 | 5 | ten-thousandths |

1.   Decimal notation is a base-10 notation for expressing _____.

2.   The name of the next place value to the right of the thousands place is the
_____ place.

3.   To determine which of two positive decimal numbers is greater, compare the digits in
the corresponding places from _____ to _____.

4.   To round a number to a given place value, round down if the digit to the right is
_____ or less.

GUIDED EXAMPLES AND PRACTICE

Objective 1   Write decimals as fractions or mixed numbers.

Review this example for Objective 1:

1.   Write as a fraction or mixed number in lowest terms.
0.4

$$0.4 = \frac{4}{10}$$ Write the decimal in the numberator.

$$= \frac{4 \div 2}{10 \div 2}$$ Simplify to lowest terms.

$$= \frac{2}{5}$$

Practice this exercise:

1.   Write as a fraction or mixed number in lowest terms.
0.35

Objective 2 Write a word name for a decimal number.

| **Review these examples for Objective 2:** | **Practice these exercises:** |

Review these examples for Objective 2:

2.   Write the word name.
      0.204

To write the word name for a decimal number
with no integer part (no digits other than 0 to
the left of the decimal point):
1. Write the word name for digits as if they
represented a whole number.
2. Write the name of the last place value.

So .204 is written as two hundred four
thousandths.

3.   Write the word name.
      10.285

To write a word name for a decimal number
with both integer and fractional parts:
1. Write the name of the integer number part.
2. Write the word *and* for the decimal point.
3. Write the name of the fractional part.

So 10.285 is written as ten and two
hundred eighty-five thousandths.

Practice these exercises:

2.   Write the word name.
      0.105

3.   Write the word name.
      125.37

Objective 3 Graph decimals on a number line.

Review this example for Objective 3:

4.   Graph on a number line.
      0.6

Because 0.6 means $\frac{6}{10}$, we divide the distance
between 0 and 1 into 10 equal-size divisions
and draw a dot on the sixth division mark.

Practice this exercise:

4.   Graph on a number line.
      2.3

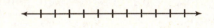

Objective 4 Use < or > to write a true statement.

Review this example for Objective 4:	**Practice this exercise:**

5. Use < or > to write a true statement.
    0.54 ? 0.5

5. Use < or > to write a true statement.
    3.457 ? 3.4569

To determine which of two positive decimal numbers is greater, compare the digits in the corresponding place values from left to right until you find two different digits in the same place. The greater number contains the greater of those two digits.

The digits match until the hundredths digit, where $4 > 0$ so $0.54 > 0.5$.

Objective 5 Round decimal numbers to a specified place.

Review this example for Objective 5:	**Practice this exercise:**

6. Round 21.06281 to the tenths place.

6. Round 21.06281 to the thousandths place.

The nearest tenths are 21.0 and 21.1. Rounding to the nearest tenth means that we consider the digit in the hundredths place, which is 6. Because 6 is greater than 5, we round up to 21.1.

ADDITIONAL EXERCISES

Objective 1 Write decimals as fractions or mixed numbers.

For extra help, see Example 1 on page 378 of your text and the Section 6.1 lecture video.
Write as a fraction or mixed number in lowest terms.

1. 0.875

2. 0.06

3. 12.2

4. −0.006

Objective 2 Write a word name for a decimal number.
For extra help, see Example 2 on page 379 of your text and the Section 6.1 lecture video.
Write the word name.

5. .00789

6. 50.0117

7. −4.83

8. −526.3247

Objective 3 Graph decimals on a number line.
For extra help, see Example 3 on page 380 of your text and the Section 6.1 lecture video.
Graph on a number line.

9. 6.25

10. 7.08

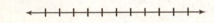

11. −5.67

12. −13.794

Objective 4 Use < or > to write a true statement.
For extra help, see Example 4 on page 381 of your text and the Section 6.1 lecture video.
Use < or > to write a true statement.

13. 541.11 ? 541.101

14. 0.003247 ? 0.003249

15. −2.3054 ? −2.0345

16. −56.94103 ? −56.94102

Objective 5 Round decimal numbers to a specified place.
For extra help, see Example 5 on page 382 of your text and the Section 6.1 lecture video.
Round 21.06281 to the specified place.

17. ten-thousandths

18. hundredths

19. whole number

20. tens

Chapter 6 DECIMALS

6.2 Adding and Subtracting Decimal Numbers

Learning Objectives
1 Add decimal numbers.
2 Subtract decimal numbers.
3 Add and subtract signed decimal numbers.
4 Simplify, add, or subtract polynomials containing decimal numbers.
5 Solve equations using the addition principle.
6 Solve applications.

GUIDED EXAMPLES AND PRACTICE

Objective 1 Add decimal numbers.

Review this example for Objective 1:
1. Add $0.42 + 3.121$

$$
\begin{array}{r}
0.42 \\
+3.121 \\
\hline
3.541
\end{array}
$$
     Line up the decimal points.

Practice this exercise:
1. Add
     $34.5289 + 96.2204$

Objective 2 Subtract decimal numbers.

Review this example for Objective 2:
2. Subtract $19.34 - 15.2$

$$
\begin{array}{r}
19.34 \\
-15.20 \\
\hline
4.14
\end{array}
$$
     Line up the decimal points.
     Add 0's as placeholders.

Practice this exercise:
2. Subtract
     $33.92 - 9.83$

Objective 3 Add and subtract signed decimal numbers.

Review this example for Objective 3:	**Practice this exercise:**

Review this example for Objective 3:

3.   Add or subtract $9.591 + (-35.36)$

Stack the number with the greater absolute value
on top and align the decimal points

$$\overset{\scriptstyle 2\ \ 14\ \ 12\ \ 15\ \ 10}{\cancel{3}\cancel{5}.\cancel{3}\cancel{6}\cancel{0}}$$
$$-\ \ 9.591$$
$$\overline{25.769}$$

Since -35.36 has the higher absolute value, the
solution is -25.769.

Practice this exercise:

3. Add or subtract
$$-38.7 - 9.806$$

Objective 4 Simplify, add, or subtract polynomials containing decimal numbers.

Review these examples for Objective 4:

4.   Simplify $6 + 9.8b + 2.7 - 6b + 7.3b - 3$

Combine like terms
$9.8b - 6b + 7.3b + 6 + 2.7 - 3$
$\quad 11.1b \qquad + \quad 5.7$

5.   Add or subtract
$$\left(0.05x^5 - 0.3x^2 + x + 0.05\right) +$$
$$\left(-0.03x^5 + x^3 - 0.3x - 0.02\right)$$

Combine like terms
$0.05x^5 - 0.03x^5 + x^3 - 0.3x^2 + x - 0.3x + 0.05 - 0.02$
$\quad 0.02x^5 \qquad + x^3 - 0.3x^2 \quad + 0.7x \qquad + 0.03$

Practice these exercises:

4.   Simplify
$$3 + 7.7d + 4.4 - 3d + 7.7d - 2$$

5.   Add or subtract
$$\left(2.39a^4 - 0.039a^3 + 1.34\right) -$$
$$\left(1.05a^4 - 0.095a^3 + 0.58a^2\right)$$

Objective 5 Solve equations using the addition principle.

Review this example for Objective 5:

6.   Solve and check $81.36 + r = 978.23$

$$81.36 + r = 978.23$$
$$\underline{-81.36 \qquad -81.36} \quad \text{Subtract } 81.36.$$
$$r = 896.87$$

Check:

$$81.36 + 896.87 \overset{?}{=} 978.23$$
$$978.23 = 978.23 \quad \text{True, so } 896.87$$
$$\text{is the solution.}$$

Practice this exercise:

6. $x - 31.91 = -47.6$

Objective 6 Solve applications.

Review this example for Objective 6:

7.   A businesswoman has $1527.32 in her checking account. She writes checks of $23.03, $590.99, and $63.79 to pay some bills. She then deposits a paycheck of $524.15. How much is in her account after these changes?

Subtract the sum of all the checks (debits) from the initial balance and add the deposit (credit).

Balance = initial balance − sum of all debts + credit

$B = 1527.32 - (23.03 + 590.99 + 63.79) + 524.15$

$B = 1527.32 - 677.81 + 524.15$

$B = 849.51 + 524.15$

$B = 1373.66$

The final balance is $1373.66.

Practice this exercise:

7.   The perimeter of the triangle shown is 65.9 millimeters. Find the length of the missing side.

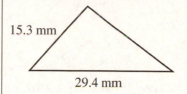

15.3 mm

29.4 mm

ADDITIONAL EXERCISES
Objective 1 Add decimal numbers.
For extra help, see Example 1 on pages 385–386 of your text and the Section 6.2 lecture video.
Add.

1. $65.72 + 12.098$

2. $5 + 81.12 + 34.0854$

3. $94.1903 + 73.4758$

4. $87 + 99.63 + 2.206$

Objective 2 Subtract decimal numbers.
For extra help, see Example 2 on pages 386–387 of your text and the Section 6.2 lecture video.
Subtract.

5. $79.84 - 1.63$

6. $56.1254 - 12.5891$

7. $7 - 3.529$

8. $95.6742 - 68.6783$

Objective 3 Add and subtract signed decimal numbers.
For extra help, see Examples 3–4 on pages 387–388 of your text and the Section 6.2 lecture video.
Add or subtract.

9. $-94 + (-40.3)$

10. $71.2 + (-21.5)$

11. $1.09 - (-12.37)$

12. $-13.9 - (-21.5)$

Objective 4 Simplify, add, or subtract polynomials containing decimal numbers.
For extra help, see Examples 5–6 on pages 388–389 of your text and the Section 6.2 lecture video.
Simplify.

13. $4.8 + n - 2.6 - 5.1n + 8.4n + 1$

14. $8.2y - 3.7y^2 - 2.3y + 45 + 6.4y - 3.6$

Add or subtract.

15. $\begin{array}{l}(0.06x^5 - 0.02x^2 + x + 0.05) + \\ (-0.04x^5 + x^3 - 0.5x - 0.04)\end{array}$

16. $\begin{array}{l}(2.77r^4 - 0.037r^3 + 1.87) - \\ (0.90r^4 - 0.059r^3 + 0.24r^2)\end{array}$

Objective 5 Solve equations using the addition principle.
For extra help, see Example 7 on page 389 of your text and the Section 6.2 lecture video.
Solve and check

17. $a + 60.73 = -48.432$

18. $b - (-51.9) = 9.34$

Objective 6 Solve applications
For extra help, see Example 8 on page 390 of your text and the Section 6.2 lecture video.

19. A business man has $1628.67 is his checking account. He writes checks of $623.11, $34.56, $63.98, and $871.47 to pay some bills. He then deposits a paycheck of $721.68. How much is in his account after these changes?

20. A triangle has a perimeter of 20 cm. Two of the sides are 6.3 cm and 9.1 cm. What is the length of the third side?

Chapter 6 DECIMALS

6.3 Multiplying Decimal Numbers; Exponents with Decimal Bases

---

Learning Objectives
1 Multiply decimal numbers.
2 Multiply signed decimal numbers.
3 Evaluate exponential forms with decimal bases.
4 Write a number in scientific notation in standard form.
5 Write standard form numbers in scientific notation.
6 Multiply monomials.
7 Multiply polynomials.
8 Solve applications.

GUIDED EXAMPLES AND PRACTICE

Objective 1 Multiply decimal numbers.

Review this example for Objective 1:

1. Multiply $(5.4)(4.3)$

$$
\begin{array}{rl}
5.4 & \text{1 decimal place} \\
\times\, 4.3 & \underline{+1 \text{ decimal place}} \\
\hline
162 & \\
+216 & \\
\hline
23.22 & \text{2 decimal places}
\end{array}
$$

Practice this exercise:

1. Multiply $206.57(0.01)$

Objective 2 Multiply signed decimal numbers.

Review this example for Objective 2:

2. Multiply $-2.6(-2.18)$

$$
\begin{array}{rl}
2.18 & \text{2 decimal places} \\
\times 2.6 & \underline{+1 \text{ decimal place}} \\
\hline
1308 & \\
+436 & \\
\hline
5.668 & \text{3 decimal places}
\end{array}
$$

Since the two numbers have the same sign, the product is positive, so the answer is 5.668.

Practice this exercise:

2. $100(-0.16)$

Objective 3 Evaluate exponential forms with decimal bases.

Review this example for Objective 3:	Practice this exercise:
3. Evaluate $(0.6)^2$.	3. Evaluate $(-0.1)^3$.

Multiply the number of decimal places in the base by the exponent to get the number of decimal places in the product.

$$(0.6)^2 = (0.6)(0.6)$$
$$= .36$$

Objective 4 Write a number in scientific notation in standard form.

Review this example for Objective 4:	Practice this exercise:
4. Write 4×10^6 in standard form; then write the word name.	4. Write 2.67×10^{11} in standard form; then write the word name.

$$4 \times 10^6 = 4\underset{1\ 2\ 3\ 4\ 5\ 6}{000000}$$
$$= 4,000,000$$

word name: 4 million

Objective 5 Write standard form numbers in scientific notation.

Review this example for Objective 5:	Practice this exercise:
5. Write 564,500,000 in scientific notation.	5. Write 7,895,450,000 in scientific notation.

Move the decimal to express a number whose absolute value is greater than or equal to 1 but less than 10.

$$564,500,000 = 5.645 \times 10^8$$

There are 8 places between the new decimal position and the original position.

Name: Date:
Instructor: Section:

Objective 6 Multiply monomials.

Review this example for Objective 6:

6.   Multiply $\left(4.8x^2\right)\left(-1.7x^4\right)$

Multiply the coefficients and add the exponents of the like variables.

$$(4.8x^2)(-1.7x^4) = 4.8 \cdot -1.7 \cdot x^{2+4}$$
$$= -8.16x^6$$

Practice this exercise:

6.   Multiply
$$\left(-12.4m^3n\right)\left(-0.7m^6\right)$$

Objective 7 Multiply polynomials.

Review this example for Objective 7:

7.   Multiply $8.6t^2\left(5t^2 + 2.9t - 0.3\right)$

Using the distributive property, multiply each term in $5t^2 + 2.9t - 0.3$ by $8.6t^2$.

$$8.6t^2\left(5t^2 + 2.9t - 0.3\right)$$
$$= 8.6t^2 \cdot 5t^2 + 8.6t^2 \cdot 2.9t - 8.6t^2 \cdot 0.3$$
$$= 43t^4 + 24.94t^3 - 2.58t^2$$

Practice this exercise:

7.   Multiply
$$\left(0.2n - 6\right)\left(3n + 0.1\right)$$

Objective 8 Solve applications.

Review this example for Objective 8:

8.   Reginald is a real estate broker who helps people sell their homes. His fee is 0.09 times the price of the home. What is his fee for selling a $120,700 home?

Multiply the fee by the price of the home.

Total fee $= 0.09 \times 120,700$
$= 10,863$

Thus his fee is $10,683.

Practice this exercise:

8.   What is the cost, in dollars, of 13.6 gallons of lead-free gasoline at 312.9 cents per gallon?

ADDITIONAL EXERCISES

Objective 1 Multiply decimal numbers.

For extra help, see Example 1 on pages 395–396 of your text and the Section 6.3 lecture video.

Multiply.

1. $(54.65)(78.49)$ **2.** $(1000)(5.1248)$

3. $(3965.34)(.001)$ **4.** $(12.68)(53.94)$

Objective 2 Multiply signed decimal numbers.

For extra help, see Example 2 on page 397 of your text and the Section 6.3 lecture video.

Multiply.

5. $42.9(-0.01)$ **6.** $(-0.416)(-0.9)$

7. $-7(5.29)$ **8.** $(-3.1415)(-1000)$

Objective 3 Evaluate exponential forms with decimal bases.

For extra help, see Example 3 on page 397 of your text and the Section 6.3 lecture video.

Evaluate.

9. $(1.5)^3$ **10.** $(-5.46)^2$

11. $(0.2)^4$ **12.** $(9.72)^2$

Objective 4 Write a number in scientific notation in standard form.

For extra help, see Example 4 on page 398 of your text and the Section 6.3 lecture video.

Write each number in standard form; then write the word name.

13. 5.897×10^7 **14.** 7.3814×10^4

Objective 5 Write standard form numbers in scientific notation.
For extra help, see Example 5 on page 399 of your text and the Section 6.3 lecture video.
Write each number in scientific notation.

15. 301,571,500,000

16. 3,600,000

Objective 6 Multiply monomials.
For extra help, see Examples 6–7 on page 400 of your text and the Section 6.3 lecture video.
Multiply.

17. $\left(3.4m^2n\right)\left(5.9mn^2\right)$

18. $\left(-0.2u^4\right)^3$

19. $\left(5.24r^3\right)\left(8.1s\right)$

20. $\left(-3x^5\right)\left(-6.45xy^2\right)$

Objective 7 Multiply polynomials.
For extra help, see Example 8 on pages 400–401 of your text and the Section 6.3 lecture video.
Multiply.

21. $0.9x^2y\left(6x^4+4.7xy-0.2\right)$

22. $2.2x^2\left(3.4x-8.2xy+6.9y^2\right)$

23. $\left(5x-4.1\right)\left(0.2x-6.5\right)$

24. $\left(5.2x-0.5\right)\left(2.8x-10\right)$

Objective 8 Solve applications.
For extra help, see Example 9 on page 402 of your text and the Section 6.3 lecture video.
Solve.

25. Carol wants to build a square deck in her backyard. Find the area of the deck if the length of one side is 3.8 meters.

26. Find the area.

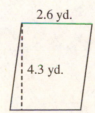

2.6 yd.

4.3 yd.

Chapter 6 DECIMALS

6.4 Dividing Decimal Numbers; Square Roots with Decimals

Learning Objectives
1   Divide decimal numbers.
2   Write fractions and mixed numbers as decimal numbers.
3   Evaluate square roots.
4   Divide monomials with decimal coefficients.
5   Solve equations using the multiplication principle.
6   Solve applications.

Key Terms
Use the vocabulary terms listed below to complete each statement in Exercises 1–4.

    rational numbers   **real numbers**   **integers**
    irrational numbers   **whole numbers**   **decimal numbers**

1.   An irrational number cannot be expressed in the form $\dfrac{a}{b}$, where a and b are

      _____ and $b \neq 0$.

2.   The _____ and the _____ together form the set of real
      numbers.

3.   To divide a decimal number by a whole number, divide the divisor into the dividend as if
      both numbers were _____.

4.   All non-terminating _____ with repeating digits can be expressed as
      fractions.

          Copyright © 2013 Pearson Education, Inc.

GUIDED EXAMPLES AND PRACTICE

Objective 1 Divide decimal numbers.

Review these examples for Objective 1:	**Practice these exercises:**

1. Divide $38.4 \div 16$

$$
\begin{array}{r}
2.4 \\
16\overline{)38.4} \\
-32 \\
\hline
64 \\
-64 \\
\hline
0
\end{array}
$$

1. Divide $0.45 \div 30$

2. Divide $-18.6 \div -0.6$

$0.6\overline{)18.6} = 6\overline{)186}$ Move the decimal 1 place to the right in both numbers.

$$
\begin{array}{r}
31 \\
6\overline{)186} \\
-18 \\
\hline
06 \\
-6 \\
\hline
0
\end{array}
$$

Since both numbers share the same sign, the quotient is positive.

2. Divide $0.01 \div 0.0005$

Objective 2 Write fractions and mixed numbers as decimal numbers.

Review this example for Objective 2:

3. Write $7\frac{3}{4}$ as a decimal number.

7 is written to the left of the decimal point. The decimal equivalent of $\frac{3}{4}$ is written to the right of the decimal.

$$
\begin{array}{r}
.75 \\
4\overline{)3.00} \\
-28 \\
\hline
20 \\
-20 \\
\hline
0
\end{array}
$$

So the answer is 7.75.

Practice this exercise:

3. Write $\frac{5}{6}$ as a decimal number.

Objective 3 Evaluate square roots.

Review this example for Objective 3:

4. Evaluate the square root $\sqrt{0.16}$. If the root is irrational, approximate the square root to the nearest hundredth.

16 is a perfect square, and 0.16 has an even number of decimal places; so its square root has half the number of decimal places and the same digits as the square root of 16, which is 4.

.4

Practice this exercise:

4. Evaluate the square root $\sqrt{350}$. If the root is irrational, approximate the square root to the nearest hundredth.

Objective 4 Divide monomials with decimal coefficients.

Review this example for Objective 4:

5. Divide $18.9x^9 \div 10.5x^7$

Divide the coefficients and subtract exponents for the like bases.

$$18.9x^9 \div 10.5x^7 = (18.9 \div 10.5) x^{9-7}$$
$$= 1.8x^2$$

Practice this exercise:

5. Divide $15.4x^7 \div 5.5x^3$

Objective 5 Solve equations using the multiplication principle.

Review this example for Objective 5:

6. Solve and check. $13.15x = 21.04$

$$\frac{13.15x}{13.15} = \frac{21.04}{13.15} \quad \text{Divide both sides by 13.15}$$
$$1x = 1.6$$
$$x = 1.6$$

Check

$$13.15x = 21.04$$

$$13.15(1.6) \overset{?}{=} 21.04 \quad \text{Replace } x \text{ with 1.6.}$$
$$21.04 = 21.04 \quad \text{True; so 1.6 is the solution.}$$

Practice this exercise:

6. Solve and check.
$$-0.75b = 6.18$$

Objective 6 Solve applications.

Review this example for Objective 6:

7. A car loan of $13,661.28 is to be paid off in 48 monthly payments. How much is each payment?

To calculate the amount of each payment (P), divide $13,661.28 by 48

$$P = 13,661.28 \div 48$$
$$P = 284.61$$

Each payment is $284.61

Practice this exercise:

7. To protect an elm tree in your backyard, you need gypsy moth caterpillar tape around the trunk. The tree has a 1.1-foot diameter. What length of tape is needed? Use 3.14 to approximate π.

ADDITIONAL EXERCISES

Objective 1 Divide decimal numbers.

For extra help, see Examples 1–3 on pages 407–408 of your text and the Section 6.4 lecture video.

Divide.

1. $6.936 \div 17$

2. $7560 \div 3.6$

3. $18,468 \div 5.7$

4. $-45.05 \div -0.05$

5. $1 \div 0.0004$

6. $125.5228 \div (-0.202)$

Objective 2 Write fractions and mixed numbers as decimal numbers.

For extra help, see Examples 4–5 on page 409 of your text and the Section 6.4 lecture video.

Write as a decimal number.

7. $14\dfrac{1}{2}$

8. $\dfrac{2}{3}$

9. $-5\dfrac{2}{5}$

10. $-4\dfrac{5}{6}$

Objective 3 Evaluate square roots.

For extra help, see Examples 6–7 on pages 410–411 of your text and the Section 6.4 lecture video.

Evaluate the square root. If the root is irrational, approximate the square root to the nearest hundredth.

11. $\sqrt{0.0625}$

12. $\sqrt{51}$

Objective 4 Divide monomials with decimal coefficients.
For extra help, see Example 8 on page 412 of your text and the Section 6.4 lecture video.
Divide.

13. $4.5x^4 \div 0.9x^3$

14. $\dfrac{34.1x^9}{-15.5x^3}$

Objective 5 Solve equations using the multiplication principle.
For extra help, see Example 9 on page 413 of your text and the Section 6.4 lecture video.
Solve and check.

15. $0.9x = 6.3$

16. $6 = 7.5y$

17. $8.7z = -59.595$

18. $-65.038 = -12.4t$

Objective 6 Solve applications.
For extra help, see Examples 10–11 on pages 413–414 of your text and the Section 6.4 lecture video.
Solve.

19. You buy 5.25 pounds of potatoes at $1.50 a pound. What is the total cost of the potatoes?

20. A car loan of $16,257.96 is to be paid off in 36 monthly payments. How much is each payment?

Chapter 6 DECIMALS

6.5 Order of Operations and Applications in Geometry

Learning Objectives
1   Simplify numerical expressions using the order of operations agreement.
2   Simplify expressions containing fractions and decimals.
3   Solve application problems requiring more than one operation.
4   Find weighted means.
5   Evaluate expressions.
6   Find the area of a triangle, trapezoid, and circle.
7   Find the volume of a cylinder, pyramid, cone, and sphere.
8   Find the area and volume of a composite form.

GUIDED EXAMPLES AND PRACTICE

Objective 1 Simplify numerical expressions using the order of operations agreement.

Review this example for Objective 1:

1.   Simplify $18 \times (47.3 + 92.2)$

$18 \times (47.3 + 92.2) = 18 \times (139.5)$ Add in the parentheses.

$= 2511$ Multiply.

Practice this exercise:

1.   Simplify

$0.25 + 5.8(0.3)$

Objective 2 Simplify expressions containing fractions and decimals.

Review this example for Objective 2:

2.   Simplify $\dfrac{1}{2}(-0.66)$

$\dfrac{1}{2}(-0.66) = \dfrac{1}{2} \cdot \dfrac{-66}{100}$ Write -0.66 as $\dfrac{-66}{100}$.

$= \dfrac{1}{\overset{}{\underset{1}{\cancel{2}}}} \cdot \dfrac{\overset{-33}{\cancel{-66}}}{100}$ Divide out common factors.

$= \dfrac{-33}{100}$, or -0.33 Multiply.

Practice this exercise:

2.   Simplify $\dfrac{1}{12} + \dfrac{3}{4}(0.6)$

Objective 3 Solve application problems requiring more than one operation.

Review this example for Objective 3:

3. The Perla family plans a vacation using a travel agent. The components of their trip planned by the travel agent are listed in the table. If they want to make 5 equal payments to cover the costs for these plans, how much will each payment be?

Description	Amount
Flights	$1444.61
Hotel	$1323.72
Rental car	$347.78
Theme park tickets	$1932.89

Add all the amounts together and divide by 5 to determine how much each payment will be.

$$P = (1444.61 + 1323.72 + 347.78 + 1932.89) \div 5$$
$$P = 5049 \div 5$$
$$P = 1009.80$$

Each payment will be $1009.80.

Practice this exercise:

3. Jonathan is traveling for business using his own car. His employer will reimburse him $0.41 per mile driven and cover all costs for meals and hotel. If he drives 372 miles and presents receipts for $72.19 for meals and $355.62 for hotel charges, how much will his employer reimburse him?

Objective 4 Find weighted means.

Review this example for Objective 4:

4. Calculate the GPA for the student whose grade reports are given.

Course	Credits	Grade
MATH 203	4.0	A
ENG 211	4.0	B
ECO 214	3.0	B+
MUS 216	2.0	B+

The following list shows commonly used grade point equivalents for each letter grade.

A = 4

B+ = 3.5

B = 3

C+ = 2.5

C = 2

D+ = 1.5

D = 1

F = 0

MATH 203 $4(4.0) = 16$

ENG 211 $\quad 4(3.0) = 12$

ECO 214 $\quad 3(3.5) = 10.5$

MUS 216 $\quad 2(3.5) = 7$

$4 + 4 + 3 + 2 = 13$

$$\text{GPA} = \frac{\text{total grade points}}{\text{total credits}}$$
$$= \frac{16 + 12 + 10.5 + 7}{13}$$
$$= \frac{45.5}{13}$$
$$= 3.5$$

Practice this exercise:

4. Calculate the GPA for the student whose grade reports are given.

Course	Credits	Grade
FRE 300	4.0	C
MATH 305	4.0	B
PHY 250	4.0	B
ENG 300	3.0	B

Objective 5 Evaluate expressions.

Review this example for Objective 5:	**Practice this exercise:**

5. Evaluate the expression using the given values. $\frac{1}{2}mv^2$; $m = 4.25$, $v = -12$

5. Evaluate the expression using the given values.
mc^2; $m = 4.7 \times 10^4$, $c = 3 \times 10^8$

$\frac{1}{2}(4.25)(-12)^2$ Replace m with 4.25, v with -12.

$0.5(4.25)(-12)^2$ Write $\frac{1}{2}$ as 0.5.

$0.5(4.25)(144)$ Evalute exponential.

306 Multiply each number together.

Objective 6 Find the area of a triangle, trapezoid, and circle.

Review these examples for Objective 6:

Practice these exercises:

6. Find the area of the triangle shown.

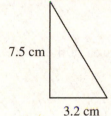

7.5 cm

3.2 cm

6. Find the area of the triangle shown.

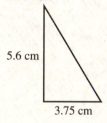

5.6 cm

3.75 cm

$A = \frac{1}{2}bh$

$A = \frac{1}{2}(3.2)(7.5)$ Replace b with 3.2 and h with 7.5

$A = \frac{1}{2}(24)$ Multiply 3.2 and 7.5

$A = \frac{24}{2}$ Multiply.

$A = 12$ Divide 24 by 2.

Therefore, the area is 12 cm.

7. Find the area of the trapezoid shown.

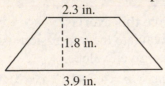

$$A = \frac{1}{2}h(a+b)$$

$$A = \frac{1}{2}(1.8)(2.3+3.9) \quad \text{Replace h, a, and b.}$$

$$A = \frac{1}{2}(1.8)(6.2) \quad \text{Add 2.3 and 3.9.}$$

$$A = \frac{1}{2}(11.16) \quad \text{Multiply 1.8 and 6.2.}$$

$$A = \frac{11.16}{2} \quad \text{Multiply.}$$

$$A = 5.58 \quad \text{Divide 11.16 by 2.}$$

Therefore, the area is 5.58 in.

8. Find the area of the circle shown. Use 3.14 for π.

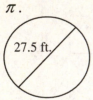

$$r = \frac{27.5}{2}$$

$$r = 13.75$$

$$A = \pi r^2$$

$$A \approx (3.14)(13.75)^2 \quad \text{Replace } r \text{ with 13.75.}$$

$$A \approx (3.14)(189.0625) \quad \text{Square 13.75.}$$

$$A \approx (593.65625) \quad \text{Multiply 13.75 and 3.14.}$$

Therefore, the area is approximately 593. 66 ft^2 .

7. Find the area of the trapezoid shown.

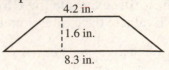

8. Find the area of the circle shown. Use 3.14 for π

Name: Date:

Instructor: Section:

Objective 7 Find the volume of a cylinder, pyramid, cone, and sphere.

Review these examples for Objective 7:

9. Find the volume. Use 3.14 for π.

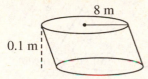

$V = \pi r^2 h$

$V \approx (3.14)(8)^2 (0.1)$ Replace the values known.

$V \approx (3.14)(64)(0.1)$ Square 8.

$V \approx (3.14)(6.4)$ Multiply 64 and 0.1.

$V \approx 20.096$ Multiply 3.14 and 6.4.

Therefore, the volume is 20.096 m^3.

10. One of the ancient stone pyramids in Egypt has a square base that measures 149 meters on each side. The height is 99 meters. What is the volume of the pyramid?

$V = \dfrac{1}{3}lwh$

$V = \dfrac{1}{3}(149)(149)(99)$ Replace the values known.

$V = \dfrac{1}{3}(22,201)(99)$ Multiply 149 and 149.

$V = \dfrac{1}{3}(2,197,899)$ Multiply 22,201 and 99.

$V = \dfrac{2,197,899}{3}$ Divide by 3.

$V = 732,633$

Therefore, the volume is 732,633 m^3.

Practice these exercises:

9. Find the volume. Use 3.14 for π.

10. Find the volume.

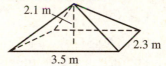

Name: Date:

Instructor: Section:

11. Find the volume. Use 3.14 for π.

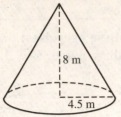

$$V = \frac{1}{3}\pi r^2 h$$

$$V \approx \frac{1}{3}(3.14)(4.5)^2(8) \quad \text{Replace with values known.}$$

$$V \approx \frac{1}{3}(3.14)(20.25)(8) \quad \text{Square 4.5.}$$

$$V \approx \frac{1}{3}(3.14)(162) \quad \text{Multiply 20.25 and 8.}$$

$$V \approx (3.14)\left(\frac{162}{3}\right) \quad \text{Divide 162 and 3.}$$

$$V \approx (3.14)(54) \quad \text{Multiply 3.14 and 54.}$$

$$V \approx 169.56$$

Therefore, the volume is approximately
169.56 m^3.

12. Find the volume. Use 3.14 for π.

$$V = \frac{4}{3}\pi r^3$$

$$V \approx \frac{4}{3}(3.14)(7.2)^3 \quad \text{Replace the values known.}$$

$$V \approx \frac{4}{3}(3.14)(373.248) \quad \text{Cube 7.2.}$$

$$V \approx (3.14)\left(\frac{1492.992}{3}\right) \quad \text{Multiply 373.248 and 4.}$$

$$V \approx (3.14)(497.664) \quad \text{Divide 1492.992 by 3.}$$

$$V \approx 1562.66 \quad \text{Multiply 3.14 and 497.664.}$$

Therefore, the volume is approximately
1562.66 mm^3.

11. Find the volume. Use 3.14 for π.

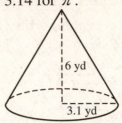

12. Find the volume. Use 3.14 for π.

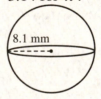

Objective 8 Find the area and volume of a composite form.

Review this example for Objective 8:

13. Find the area of the shaded region. Use 3.14 for π.

4.5 cm

4.5 cm

The figure is a half circle and a triangle, so add their areas together. Note that the radius is half of 4.5, which is 2.25. Create the area formula, replace with the values known, then simplify.

$A =$ area of half circle + area of triangle.

$$A = \quad \frac{1}{2}\pi r^2 \quad + \quad \frac{1}{2}bh$$

$$A \approx \frac{1}{2}(3.14)(2.25)^2 + \frac{1}{2}(4.5)(4.5)$$

$$A \approx \frac{1}{2}(3.14)(5.0625) + \frac{1}{2}(4.5)(4.5)$$

$$A \approx 0.5(5.0625)(3.14) + 0.5(4.5)(4.5)$$

$$A \approx 2.53125(3.14) + 10.125$$

$$A \approx 7.948125 + 10.125$$

$$A \approx 18.073125$$

The area is about 18.07 cm^2.

Practice this exercise:

13. Find the area of the shaded region. Use 3.14 for π.

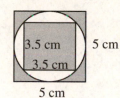

3.5 cm   5 cm

3.5 cm

5 cm

ADDITIONAL EXERCISES

Objective 1 Simplify numerical expressions using the order of operations agreement.

For extra help, see Example 1 on page 418 of your text and the Section 6.5 lecture video.

Simplify.

1. $(0.5)^2 - 5.4 \div 0.1(6.3)$

2. $\sqrt{0.0009} + 116.4 \div 4(2.5 - 5.8)$

3. $8.1 + 3.3\sqrt{0.16} - 41.2 \div 2$

4. $\left[-6.27 \div (2.93 + .07)\right] + 0.4\sqrt{1.44}$

Objective 2 Simplify expressions containing fractions and decimals.

For extra help, see Example 2 on page 419 of your text and the Section 6.5 lecture video.

Simplify.

5. $1\dfrac{1}{4}(2.5)-\dfrac{1}{8}$

6. $\dfrac{1}{6}\div(-0.5)+\left(\dfrac{1}{5}\right)^2$

7. $\left(2\dfrac{3}{4}\right)^2(4.2)-5.5625$

8. $-0.7\div\dfrac{2}{5}-4\dfrac{1}{5}$

Objective 3 Solve application problems requiring more than one operation.

For extra help, see Example 3 on pages 420–421 of your text and the Section 6.5 lecture video.

Solve.

9. A student has gotten the following scores on the past 5 tests, 90, 85, 95, 73, 86. What is the mean of these scores?

10. An employee is traveling for business using his own car. His employer will reimburse him $0.48 per mile driven and cover all costs for meals and hotel. If he drives 462 miles and presents receipts for $82.90 for meals and $471.29 for hotel charges, how much will his employer reimburse him?

Objective 4 Find weighted means.

For extra help, see Example 4 on pages 421–422 of your text and the Section 6.5 lecture video.

Calculate the GPA for the students whose grade reports are given.

11.

Course	Credits	Grade
Math 100	4	C+
ENG 101	4	A
PSY 101	3	B
BIO 101	3	B+

12.

Course	Credits	Grade
Math 400	5	B+
SPA 401	4	A
PSY 405	3	C
MUS 425	3	C+

Objective 5 Evaluate expressions.
For extra help, see Example 5 on page 422 of your text and the Section 6.5 lecture video.
Evaluate each expression using the given values.

13. $\frac{1}{2}mv^2$; $m = 3.75$, $v = -16$

14. mc^2; $m = 2.4 \times 10^5$, $c = 3 \times 10^6$

Objective 6 Find the area of a triangle, trapezoid, and circle.
For extra help, see Example 6 on pages 423–424 of your text and the Section 6.5 lecture video.
Find each area. Use 3.14 for π.

15.

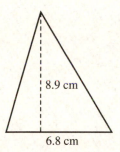

8.9 cm

6.8 cm

16.

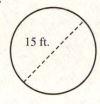

15 ft.

17.

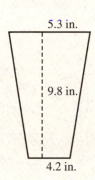

5.3 in.

9.8 in.

4.2 in.

18.

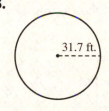

31.7 ft.

Objective 7 Find the volume of a cylinder, pyramid, cone, and sphere.
For extra help, see Examples 7–10 on pages 424–426 of your text and the Section 6.5 lecture video.
Find each volume. Use 3.14 for π.

19.

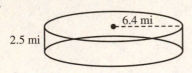

20.

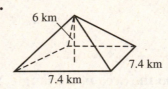

21.

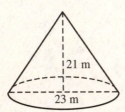

22.

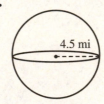

Objective 8 Find the area and volume of a composite form.
For extra help, see Examples 11–12 on pages 427–428 of your text and the Section 6.5 lecture video.
Find the area or volume of the region. Use 3.14 for π.

23.

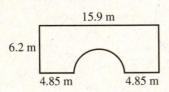

24.

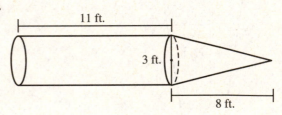

Chapter 6 DECIMALS

6.6 Solving Equations and Problem Solving

Learning Objectives
1 Solve equations using the addition and multiplication principles of equality.
2 Eliminate decimal numbers from equations using the multiplication principle.
3 Solve problems involving one unknown.
4 Solve problems using the Pythagorean theorem.
5 Solve problems involving two unknowns.

Key Terms
Use the terms listed below to complete each statement in Exercises 1–2.

      right   hypotenuse   sum   leg   difference

1. The _____ of a right triangle is opposite the 90° angle.

2. The Pythagorean theorem states that given a right triangle, the _____ of the squares of the lengths of the legs equals the square of the hypotenuse.

GUIDED EXAMPLES AND PRACTICE

Objective 1   Solve equations using the addition and multiplication principles of equality.

Review this example for Objective 1:

1. Solve and check. $4.6a + 4.17 = -27.57$

$$4.6a + 4.17 = -27.57$$

$$\underline{-4.17 \quad -4.17} \qquad \text{Subtract 4.17 from both sides.}$$

$$\frac{4.6a + 0}{4.6} = \frac{-31.74}{4.6} \qquad \text{Divide by 4.6}$$

$$a = -6.9$$

Check

$$4.6(-6.9) + 4.17 \stackrel{?}{=} -27.57$$

$$-31.74 + 4.17 \stackrel{?}{=} -27.57$$

$$-27.57 = -27.57$$

True, so -6.9 is the solution.

Practice this exercise:

1. Solve and check.
$$2.9y - 34 = 3.1y - 34.2$$

Objective 2 Eliminate decimal numbers from equations using the multiplication principle.

Review this example for Objective 2:

2. Solve and check. $2.5(x+8)=14.4+1.8x$

$$2.5(x+8)=14.4+1.8x$$

$2.5x+20=14.4+1.8x$ Distribute.

$10\cdot2.5x+10\cdot20=10\cdot14.4+10\cdot1.8x$ Multiply by 10.

$25x+200=144+18x$ Combine like terms.

$\underline{-18x \qquad\qquad -18x}$ Subtract 18x.

$7x+200= 144 +0$

$\qquad\qquad\underline{-200 \quad -200}$ Subtract 200.

$\dfrac{7x+\quad 0}{7}=\dfrac{-56}{7}$ Divide by 7.

$x=-8$

Check

$2.5(-8+8)\overset{?}{=}14.4+1.8(-8)$

$2.5\cdot0\overset{?}{=}14.4-14.4$

$0=0$

True, so -8 is the solution.

Practice this exercise:

2. Solve and check.

$1.5(x+4)=5.7+0.6x$

Objective 3 Solve problems involving one unknown.

Review this example for Objective 3:

3. Solve. 1.68 plus 7.7 times p is equal to 8.9 times p.

Practice this exercise:

3. Solve. 4 times h decreased by 8.125 is the product of h and -2.5

$$\underbrace{1.68 \text{ plus } 7.7 \text{ times } p}_{\downarrow} \quad \underbrace{\text{is equal to}}_{\downarrow} \quad \underbrace{8.9 \text{ times } p}_{\downarrow}$$

$$\overbrace{1.68 + 7.7p} \qquad \overbrace{=} \qquad \overbrace{8.9p}$$

$100 \cdot 1.68 + 100 \cdot 7.7p = 100 \cdot 8.9p$ Multiply by 100.

$$\begin{array}{rcl} 168 + \;\; 770p & = & 890p \\ -770p & & -770p \end{array}$$ Subtract 770p.

$$\dfrac{168 + \quad 0}{120} = \dfrac{120\,p}{120}$$ Divide by 120.

$$1.4 = p$$

Objective 4 Solve problems using the Pythagorean theorem.

Review this example for Objective 4:

4.   Find the length of the missing side.

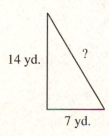

14 yd.   ?

7 yd.

$$a^2 + b^2 = c^2$$

$(7)^2 + (14)^2 = c^2$ Replace with values known.

$49 + 196 = c^2$ Square 7 and 14.

$245 = c^2$ Add 49 and 196.

$\sqrt{245} = c$ Take the square root of both sides.

$15.7 \approx c$

The length of the missing side is about 15.7 yd.

Practice this exercise:

4.   A slow-pitch softball diamond is actually a square 66 feet on a side. How far is it from home to second base?

66 ft.

Objective 5 Solve problems involving two unknowns.

Review this example for Objective 5:

5. A cashier has a total of 25 coins, made up of dimes and half dollars. The total value of the money is $6.90. How many of each kind does he have?

Half dollars + Dimes = amount of money

$$.50n \quad +.10(25-n) = 6.90$$

$.50n + 2.5 - .10n = 6.90$	Distribute the .10.
$10 \cdot .50n + 10 \cdot 2.5 - 10 \cdot .10n = 10 \cdot 6.90$	Multiply by 10.
$5n + 25 - n = 69$	Simplify.
$4n + 25 = 69$	Combine like terms.
$\underline{-25 \quad -25}$	Subtract 25.
$\dfrac{4n}{4} + 0 = \dfrac{44}{4}$	Divide by 4.

$n = 11$ Number of half dollars.

$25 - 11 = 14$ Number of dimes.

Practice this exercise:

5. Alex purchased CDs for $8.99 and DVDs for $14.99 as gifts. If he bought 6 more CDs than DVDs and spent a total of $149.86 before sales tax, how many of each did he purchase?

ADDITIONAL EXERCISES

Objective 1   Solve equations using the addition and multiplication principles of equality.

For extra help, see Example 1 on page 436 of your text and the Section 6.6 lecture video.

Solve and check.

1. $6.1y - 8 = -5.2 + 8.9y$

2. $4.2x - 0.44 = 0.84 - 2.2x$

3. $5.3x - 10.5 = 7.95 - 6.7x$

4. $10.5y - 94.645 = -64.825 + 3.4y$

Objective 2   Eliminate decimal numbers from equations using the multiplication principle.

For extra help, see Examples 2–3 on pages 437–438 of your text and the Section 6.6 lecture video.

Solve and check.

5. $3m - 1.49 = 1.4(m-1)$

6. $7(2.75 - x) - 5.4x = 8.9 - (4 + 6.1x)$

Objective 3   Solve problems involving one unknown.

For extra help, see Examples 4–5 on pages 438–439 of your text and the Section 6.6 lecture video.

Solve.

7.   64.6 less than 8 times n is equal to 200 plus 1.7 times n.

8.   Tommy uses a cell phone service that charges $59.99 per month for 550 minutes. After 550 minutes, it costs $0.55 for each additional minute. If Tommy's bill is $117.74, how many total minutes did he use?

Objective 4   Solve problems using the Pythagorean theorem.

For extra help, see Example 6 on pages 441–442 of your text and the Section 6.6 lecture video.

Solve.

9.   Holly's new cordless telephone has clear reception up to 190 feet from its base. Her phone base is located near a window in her apartment, 114 feet above street level. How far into the backyard can Holly use her phone?

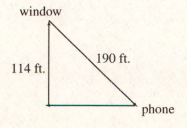

10.   A baseball diamond is actually a square 27.4 meters on a side. How far is it from home to second base?

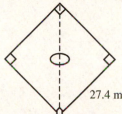

27.4 m

Objective 5 Solve problems involving two unknowns.
For extra help, see Example 7 on pages 443–444 of your text and the Section 6.6
lecture video.
Solve.

11. An ice cream vendor sells two
sizes of ice cream cones, small and
large. One day he sold 28 more
small cones than large cones for a
total of $78.50. If a small cone
costs $1.25 and a large cone costs
$1.65, how many of each size did
he sell that day?

12. There are 35 coins in your pocket, all
nickels and quarters. You have $2.55 in
your pocket. How many of each coin do
you have?

Chapter 7 RATIOS, PROPORTIONS, AND MEASUREMENT

7.1 Ratios, Probability, and Rates

Learning Objectives
1 Write ratios in simplest form.
2 Determine probabilities.
3 Calculate unit ratios.
4 Write rates in simplest form.
5 Use unit price to determine the better buy.

Key Terms

Use the terms listed below to complete each statement in Exercises 1–4.

possible   rate   ratio   product   quotient   favorable   unit

1. A ratio comparing two different measurements is called a(n) _____.

2. When the denominator is 1, the ratio is called a(n) _____ ratio.

3. The theoretical probability of equally likely outcomes is the ratio of the number of
_____ outcomes to the number of _____ outcomes.

4. A ratio is a comparison of two quantities using a(n) _____.

GUIDED EXAMPLES AND PRACTICE

Objective 1 Write ratios in simplest form.

Review this example for Objective 1:
1. Write each ratio in simplest form. There are 36 males and 39 females in a school marching band.
    a. What is the ratio of female band members to total band members?
    b. What is the ratio of male band members to female band members?

Practice this exercise:
1. Write each ratio in simplest form. A theater sells tickets at different price levels depending on the location of the seats. The following table shows the number of seats sold at each price level for a particular show.

Price	Number of Seats
$80	875
$65	180
$45	375

$$\frac{\text{female band members}}{\text{total band members}}$$

$\dfrac{39}{39+36}$ Replace the values known.

$\dfrac{39}{75}$ Add 39 and 36.

$\dfrac{13}{25}$ Divide the numerator and denominator by 3.

b.   The ratio of male band members to female band member is the following:

$$\frac{\text{male band members}}{\text{female band members}}$$

$\dfrac{36}{39}$ Replace the values known.

$\dfrac{12}{13}$ Divide the numerator and denominator by 3.

a. What is the ratio of the number of $80 seats to the number of $45 seats?

b. What is the ratio of the number of $65 seats to the total number of seats?

Objective 2 Determine probabilities.

Review this example for Objective 2:

2.   What is the probability of selecting a 10 in a standard deck of cards? Write the result in simplest form.

$$\frac{\text{Number of tens in deck}}{\text{Total cards in deck}}$$

$\dfrac{4}{52}$ There are 4 tens, one of each suit.

$\dfrac{1}{13}$ Divide the numerator and denominator by 4.

Practice this exercise:

2.   What is the probability of selecting a spade in a standard deck of cards? Write the result in simplest form.

Objective 3 Calculate unit ratios.

Review this example for Objective 3:

3. A family has a total debt of $5923 and a gross income of $25,525. What is their debt-to-income ratio? Interpret the result.

$$\text{Debt-to-income ratio} = \frac{5923}{25,525}$$

$$\approx 0.23$$

The family's debt to income ratio is approximately 0.23, which means it owes about $0.23 for every $1 of income.

Practice this exercise:

3. The stock of Pest B Gone has a current share price of $67\frac{9}{16}$ and annual net earnings per share of $3.89. What is the ratio of selling price to annual earning per share? Interpret the result.

Objective 4 Write rates in simplest form.

Review this example for Objective 4:

4. A long-distance phone call between two cities costs $2.21 for 17 minutes. What is the unit price in dollars per minute?

Write the number of dollars in the numerator and the number of minutes in the denominator.

$$\frac{\$2.21}{17 \text{ minutes}}$$

$$\frac{\$0.13}{1 \text{ minute}} \quad \text{Divide 2.21 and 17.}$$

$0.13 / \text{minute}$

Practice this exercise:

4. A $\frac{1}{3}$-pound package of cheese costs $0.59. What is the unit price in dollars per pound?

Objective 5 Use unit price to determine the better buy.

Review this example for Objective 5:

5. Determine the better buy: a 3.9-ounce package of instant pudding for $0.64 or a 5.9-ounce package of instant pudding for $0.82

Practice this exercise:

5. Determine the better buy: 3 bars of brand A soap for $1.11 or 4 bars of brand B soap for $1.75

$$\frac{\$0.64}{3.9 \text{ oz.}}$$

$0.1641 / \text{oz}$

versus

$$\frac{\$0.82}{5.9 \text{ oz.}}$$

$0.1390 / \text{oz.}$

Because $0.14 is the smaller unit price, buying the 5.9-ounce package is the better buy.

ADDITIONAL EXERCISES

Objective 1 Write ratios in simplest form.

For extra help, see Examples 1–3 on pages 463–464 of your text and the Section 7.1 lecture video.

Write each ratio in simplest form.

1. The back wheel of a bicycle rotates $1\frac{3}{7}$ times with $2\frac{2}{5}$ rotations of the pedals. Write the ratio of back wheel rotations to pedal rotations in simplest form.

2. The following table shows the number of each type of song stored on an MP3 player.

Genre	Number of Songs
Rock	1650
Rap	54
Pop	1320
Country	16

a. What is the ratio of rock songs to rap songs?

b. What is the ratio of pop songs to rock songs?

c. What is the ratio of country songs to total songs?

Objective 2 Determine probabilities.

For extra help, see Example 4 on page 466 of your text and the Section 7.1 lecture video.

Determine each probability. Write the result in simplest form.

3. What is the probability of selecting a 7 or a Jack in a standard deck of cards?

4. What is the probability of selecting a black queen in a standard deck of cards?

5. What is the probability of tossing an odd number on a six-sided game die?

6. A box of gumdrops was found to contain the number of gumdrops listed in the table.

Type	Number
Lemon	6
Orange	2
Cherry	6
Grape	6

a. If one gumdrop is drawn from the box, what is the probability of drawing an orange?

b. What is the probability of drawing a blueberry?

Objective 3 Calculate unit ratios.

For extra help, see Example 5 on page 467 of your text and the Section 7.1 lecture video.

Calculate each ratio as a unit ratio and interpret the result.

7. A family has a total debt of $5473 and a gross income of $13,031. What is their debt-to-income ratio? Interpret the result.

8. A community college has 12,315 students and 730 faculty. What is the student-to-faculty ratio? Interpret the result.

Objective 4 Write rates in simplest form.

For extra help, see Examples 6–7 on pages 468–469 of your text and the Section 7.1 lecture video.

Solve.

9. A 20-ounce box of cereal costs $5.05. What is the unit price in cents per ounce?

10. Anna drove 246.5 miles in $4\dfrac{1}{4}$ hours. What was Anna's average rate in miles per hour?

Objective 5 Use unit price to determine the better buy.

For extra help, see Example 8 on page 469 of your text and the Section 7.1 lecture video.

Determine the better buy.

11. Half a pound of brand A deli meat for $1.79 or a quarter pound of brand B deli meat for $0.90

12. Three 15-ounce cans of peas for $1.26 or two 20-ounce cans of peas for $1.20

Chapter 7 RATIOS, PROPORTIONS, AND MEASUREMENT

7.2 Proportions

Learning Objectives
1   Determine whether two ratios are proportional.
2   Solve for an unknown value in a proportion.
3   Solve proportion problems.
4   Use proportions to solve for unknown lengths in similar figures.

Key Terms
Use the terms listed below to complete each statement in Exercises 1–4.

proportional   cross products   quotients   similar   congruent

1. Two figures are similar if they have _____ angles and proportional side lengths.

2. If two ratios are _____, their cross products are equal.

3. To solve a proportion, calculate the _____ and set them equal to one another.

4. Two squares are always _____.

GUIDED EXAMPLES AND PRACTICE

Objective 1 Determine whether two ratios are proportional.

Review this example for Objective 1:
1. Determine whether the ratios are proportional.

$$\frac{10.5}{2.5} = \frac{21}{5}$$

Take the cross product of the ratios.

$$\frac{10.5}{2.5} = \frac{21}{5}$$

$$(5)(10.5) \overset{?}{=} (2.5)(21)$$

$$52.5 = 52.5$$

The ratios are proportional.

Practice this exercise:
1. Determine whether the ratios are proportional.

$$\frac{7.25}{1.75} = \frac{29}{7}$$

Objective 2 Solve for an unknown value in a proportion.

Review this example for Objective 2:

2.   Solve. $\dfrac{5}{7} = \dfrac{h}{42}$

   Calculate the cross products and set them equal to each other.

$$\frac{5}{7} = \frac{h}{42}$$

$$(42)(5) = (7)(h)$$

$210 = 7h$ Multiply both sides.

$30 = h$ Divide 210 and 7.

Practice this exercise:

2.   Solve. $\dfrac{7}{8} = \dfrac{q}{40}$

Objective 3 Solve proportion problems.

Review this example for Objective 3:

3.   If a car travels 110 miles in 3 hours, how far will it go in 7 hours?

$$\frac{110 \text{ miles}}{3 \text{ hours}} = \frac{x \text{ miles}}{7 \text{ miles}} \quad \text{Set up the proportion.}$$

$(7)(110) = (3)(x)$ Cross multiply.

$770 = 3x$ Multiply.

$256.\overline{6} = x$ Divide by 3.

   In 7 hours the car will travel approximately 256.7 miles.

Practice this exercise:

3.   If it costs $2 to buy 15 apples, how many apples will $7 buy?

Objective 4 Use proportions to solve for unknown lengths in similar figures.

Review this example for Objective 4:

4.   Find the unknown length in the similar figures.

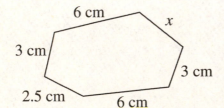

Practice this exercise:

4.   The Eiffel Tower is found to have a shadow measuring 200 meters in length. If Hugh is standing right next to the Eiffel Tower, his height is 1.2 meters and his shadow is 1.6 meters long, how tall is the Eiffel Tower?

1.5 cm
0.75 cm —
0.625 cm   1 cm
                0.75 cm
0.75 cm   1.5 cm

Choose any side to set up the proportion
to find the unknown value.

$$\frac{6}{1.5} = \frac{x}{1}$$

$$(6)(1) = (1.5)(x)$$

$$6 = 1.5x$$

$$4 = x$$

The unknown length is 4 cm.

ADDITIONAL EXERCISES
Objective 1 Determine whether two ratios are proportional.
For extra help, see Example 1 on pages 475–476 of your text and the Section 7.2
lecture video.
Determine whether the ratios are proportional.

1. $\dfrac{7}{12.2} = \dfrac{28.7}{49.8}$

2. $\dfrac{5.7}{16} = \dfrac{17.1}{49}$

3. $\dfrac{8.5}{4.02} = \dfrac{17}{8.04}$

4. $\dfrac{2\frac{1}{4}}{8\frac{1}{8}} = \dfrac{5\frac{5}{8}}{18\frac{1}{4}}$

Objective 2 Solve for an unknown value in a proportion.
For extra help, see Examples 2–3 on pages 476–477 of your text and the Section 7.2
lecture video.
Solve.

5. $\dfrac{8}{n} = \dfrac{3}{7}$

6. $\dfrac{9}{2} = \dfrac{63}{c}$

7. $\dfrac{5}{13} = \dfrac{x}{65}$

8. $\dfrac{8.5}{y} = \dfrac{17}{2}$

Objective 3 Solve proportion problems.
For extra help, see Example 4 on page 478 of your text and the Section 7.2 lecture video.
Solve.

9. Joan's family drinks 2 gallons of milk in 5 days. How many gallons does her family drink in 3 days?

10. Clarise pays $2014 in taxes on her house, which is valued at $240,000. She is planning to buy a house valued at $300,000. At the same rate, how much will her taxes be on the new house?

11. If 20 pounds of dog food costs $16.17, how much would a 30-pound bag cost?

12. On a map, 1.5 inches represents 10 miles. How many miles does 7 inches represent?

Objective 4 Use proportions to solve for unknown lengths in similar figures.
For extra help, see Examples 5–6 on pages 479–480 of your text and the Section 7.2 lecture video.
Find the unknown lengths in the similar figures.

13.

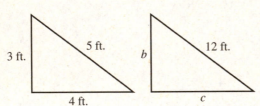

14. A skyscraper has a shadow measuring 255 meters in length. If a man is standing right next to the skyscraper, his height is 1.3 meters and his shadow is 1.7 meters long, how tall is the skyscraper?

Chapter 7 RATIOS, PROPORTIONS, AND MEASUREMENT

7.3 American Measurement; Time

Learning Objectives
1   Convert American units of length.
2   Convert American units of area.
3   Convert American units of capacity.
4   Convert American units of weight.
5   Convert units of time.
6   Convert American units of speed.

GUIDED EXAMPLES AND PRACTICE

Objective 1   Convert American units of length.

Review this example for Objective 1:

1. Convert 24 yards to feet using dimensional analysis.

$$\frac{24 \text{ yd.}}{1}$$

$$\frac{24 \text{ \cancel{yd.}}}{1} \cdot \frac{3 \text{ ft.}}{1 \text{ \cancel{yd.}}}$$ Multiply by the conversion factor.

72 ft.

Practice this exercise:

1. Convert 4 feet to inches using dimensional analysis.

Objective 2   Convert American units of area.

Review this example for Objective 2:

2. Convert 24 square yards to square feet using dimensional analysis.

$$\frac{24 \text{ yd.}^2}{1}$$

$$\frac{24 \text{ \cancel{yd.}}^2}{1} \cdot \frac{9 \text{ ft.}^2}{1 \text{ \cancel{yd.}}^2}$$ Multiply by the conversion factor.

216 ft.^{2}

Practice this exercise:

2. Convert 73 square yards to square feet using dimensional analysis.

Objective 3 Convert American units of capacity.

Review this example for Objective 3:

3. Convert 8 pints to cups using dimensional analysis.

$$\frac{8 \text{ pints}}{1}$$

$$\frac{8 \text{ pints}}{1} \cdot \frac{2 \text{ cups}}{1 \text{ pint}} \quad \text{Multiply by the conversion factor.}$$

16 cups

| **Practice this exercise:** |
| 3. Convert 12 quarts to pints using dimensional analysis. |

Objective 4 Convert American units of weight.

Review this example for Objective 4:

4. Convert 8.3 tons to pounds using dimensional analysis.

$$\frac{8.3 \text{ tons}}{1}$$

$$\frac{8.3 \text{ tons}}{1} \cdot \frac{2000 \text{ pounds}}{1 \text{ ton}} \quad \text{Multiply by the conversion factor.}$$

16,600 pounds

| **Practice this exercise:** |
| 4. Convert 4 pounds to ounces using dimensional analysis. |

Objective 5 Convert units of time.

5. Convert 16.3 minutes to seconds using dimensional analysis.

$$\frac{16.3 \text{ minutes}}{1}$$

$$\frac{16.3 \text{ minutes}}{1} \cdot \frac{60 \text{ seconds}}{1 \text{ minutes}} \quad \text{Multiply by the conversion factor.}$$

978 seconds

| **Practice this exercise:** |
| 5. Convert 300 minutes to hours using dimensional analysis. |

Objective 6 Convert American units of speed.

6. The speed of sound in dry air at 5° C is approximately 1098 feet per second. Calculate the speed of sound at 5° C in miles per minute.

$$\frac{1098 \text{ feet}}{1 \text{ second}}$$

$$\frac{1098 \text{ feet}}{1 \text{ second}} \cdot \frac{60 \text{ seconds}}{1 \text{ minute}} \quad \text{Convert seconds to minutes.}$$

$$\frac{65{,}880 \text{ feet}}{1 \text{ minute}} \cdot \frac{1 \text{ mile}}{5280 \text{ feet}} \quad \text{Convert feet to miles.}$$

$$\approx 12.477 \text{ miles / minute}$$

Practice this exercise:

6. The escape velocity for the Sun is approximately 2,025,936 feet per second. Calculate this speed in miles per hour.

ADDITIONAL EXERCISES
Objective 1 Convert American units of length.
For extra help, see Examples 1–2 on pages 485–487 of your text and the Section 7.3 lecture video.
Convert using dimensional analysis.

1. 60 inches to feet

2. 16 feet to yards

3. 7.5 miles to feet

4. 4250 feet to miles

Objective 2 Convert American units of area.
For extra help, see Examples 3–4 on pages 487–488 of your text and the Section 7.3 lecture video.
Convert using dimensional analysis.

5. 142 square inches to square yards

6. 432 square inches to square feet

Objective 3 Convert American units of capacity.
For extra help, see Example 5 on page 489 of your text and the Section 7.3 lecture video.
Convert using dimensional analysis.

7. 150 ounces to pints

8. 0.6 gallons to ounces

Objective 4 Convert American units of weight.
For extra help, see Example 6 on page 489 of your text and the Section 7.3 lecture video.
Convert using dimensional analysis.

9. $17\frac{1}{3}$ pounds to ounces

10. 6000 pounds to tons

Objective 5 Convert units of time.
For extra help, see Example 7 on page 490 of your text and the Section 7.3 lecture video.
Convert using dimensional analysis.

11. 60 days to years

12. 15 years to minutes

Objective 6 Convert American units of speed.
For extra help, see Example 8 on pages 490–491 of your text and the Section 7.3 lecture video.
Solve using dimensional analysis.

13. A car is traveling 65 miles per hour. Calculate this speed in feet per second.

14. A commercial airliner travels 807 feet per second. Calculate this speed in miles per hour.

Chapter 7 RATIOS, PROPORTIONS, AND MEASUREMENT

7.4 Metric Measurement

Learning Objectives
1 Convert units of metric length.
2 Convert units of metric capacity.
3 Convert units of metric mass.
4 Convert units of metric area.

GUIDED EXAMPLES AND PRACTICE

Objective 1 Convert units of metric length.

Review this example for Objective 1:

1. Convert 7.2 meters to centimeters.

   Centimeters are the second unit to the right of meters, so move the decimal point 2 places to the right of its position at 7.2.

   7.2 m = 720 cm

Practice this exercise:

1. Convert 0.73 meters to millimeters.

Objective 2 Convert units of metric capacity.

Review this example for Objective 2:

2. Convert 9.1 centiliters to liters.

   Liters are the second unit to the left of centiliters, so move the decimal point 2 places to the left of its position at 9.1.

   9.1 cl = 0.091 L

Practice this exercise:

2. Convert 7800 milliliters to centiliters.

Objective 3 Convert units of metric mass.

Review this example for Objective 3:

3.   Convert 800 milligrams to grams.

Grams are the third unit to the left of milligrams, so move the decimal point 3 places to the left of its position at 800.

800 mg = 0.8 g

Practice this exercise:

3.   Convert 0.8 kilograms to grams.

Objective 4 Convert units of metric area.

Review this example for Objective 4:

4.   Convert 0.022 square kilometers to square meters.

Square meters are the third unit to the right of square kilometers, so move the decimal point $3 \cdot 2 = 6$ places to the right of its position at 0.022.

$0.022 \text{ km}^2 = 22{,}000 \text{ m}^2$

Practice this exercise:

4.   Convert 38 square meters to square kilometers.

ADDITIONAL EXERCISES

Objective 1 Convert units of metric length.

For extra help, see Example 1 on pages 495–496 of your text and the Section 7.4 lecture video.

Convert.

1.   2.3 kilometers to meters

2.   125 centimeters to dekameters

3.   730 decimeters to hectometers

4.   12.5 centimeters to millimeters

5.   4 kilometers to decimeters

6.   538 centimeters to kilometers

Objective 2 Convert units of metric capacity.
For extra help, see Example 2 on page 497 of your text and the Section 7.4 lecture video.
Convert.

7. 0.01 dekaliters to kiloliters

8. 0.611 liters to milliliters

9. 6 kiloliters to liters

10. 4.19 hectoliters to liters

11. 15 cubic centimeters to milliliters

12. 7.07 cubic centimeters to liters

Objective 3 Convert units of metric mass.
For extra help, see Examples 3–4 on pages 497–498 of your text and the Section 7.4 lecture video.
Convert.

13. 0.0089 hectograms to milligrams

14. 0.052 kilograms to metric tons

15. 0.025 dekagrams to centigrams

16. 7200 grams to kilograms

17. 6 grams to decigrams

18. 3.66 metric tons to kilograms

Objective 4 Convert units of metric area.
For extra help, see Example 5 on pages 499–500 of your text and the Section 7.4 lecture video.
Convert.

19. 0.017 square decimeters to square centimeters

20. 8400 square millimeters to square meters

21. 13,000 square meters to square dekameters

22. 47,200 square hectometers to square kilometers

Chapter 7 RATIOS, PROPORTIONS, AND MEASUREMENT

7.5 Converting Between Systems; Temperature

Learning Objectives
1   Convert units of length.
2   Convert units of capacity.
3   Convert units of mass/weight.
4   Convert units of temperature.

GUIDED EXAMPLES AND PRACTICE

Objective 1 Convert units of length.

Review this example for Objective 1:

1.   The distance from the 30-yard line to the 50-yard line on a football field is 20 yards. Convert to meters.

$$\frac{20 \text{ yards}}{1}$$

$$\frac{20 \text{ yards}}{1} \cdot \frac{0.914 \text{ meters}}{1 \text{ yard}} \quad \text{Multiply by the conversion factor.}$$

18.28 meters.

Practice this exercise:

1.   The distance from the starting line of the Boston Marathon to the finish line is 138,443.5 feet. Convert to meters.

Objective 2 Convert units of capacity.

Review this example for Objective 2:

2.   How many liters are in 169.6 ounces of soda?

$$\frac{169.6 \text{ ounces}}{1}$$

$$\frac{169.6 \text{ ounces}}{1} \cdot \frac{1 \text{ cup}}{8 \text{ ounces}} \quad \text{Convert to cups.}$$

$$\frac{21.2 \text{ cups}}{1} \cdot \frac{1 \text{ pint}}{2 \text{ cups}} \quad \text{Convert to pints.}$$

$$\frac{10.6 \text{ pints}}{1} \cdot \frac{1 \text{ quart}}{2 \text{ pints}} \quad \text{Convert to quarts.}$$

$$\frac{5.3 \text{ quarts}}{1} \cdot \frac{0.946 \text{ liters}}{1 \text{ quart}} \quad \text{Convert to liters.}$$

5.0138 liters

Practice this exercise:

2.   The heavy duty wash cycle in a dishwasher uses 8.8 gallons of water. How many liters does it use?

Objective 3 Convert units of mass/weight.

Review this example for Objective 3:

3. A man weighs 262 pounds. How many kilograms is this?

$$\frac{262 \text{ pounds}}{1}$$

$$\frac{262 \text{ pounds}}{1} \cdot \frac{1 \text{ kilogram}}{2.2 \text{ pounds}} \quad \begin{array}{l}\text{Multiply by the} \\ \text{conversion factor.}\end{array}$$

119.09 kilograms

Practice this exercise:

3. A person from Europe claims that her mass is 55 kilograms. What is her weight in pounds?

Objective 4 Convert units of temperature.

Review this example for Objective 4:

4. On a cold day in December, the temperature is reported to be 3° C. What is this in degrees Fahrenheit?

Use the formula $F = \dfrac{9}{5}C + 32$ and solve.

$$F = \frac{9}{5}C + 32$$

$$F = \frac{9}{5}(3) + 32 \quad \text{Replace } C \text{ with 3.}$$

$$F = 5.4 + 32 \quad \text{Multiply 3 and } \frac{9}{5}.$$

$$F = 37.4 \quad \text{Add 5.4 and 32.}$$

Practice this exercise:

4. A city recorded a temperature of 101° F. Convert the temperature to degrees Celsius.

ADDITIONAL EXERCISES
Objective 1 Convert units of length.

For extra help, see Examples 1–2 on pages 502–503 of your text and the Section 7.5 lecture video.

Convert.

1. The width of a piece of paper is $8\dfrac{1}{2}$ inches. How many millimeters is this?

2. The height of a skyscraper is 415 meters. How many yards is this?

3. How many miles are equal to 1256.345 kilometers?

4. How many feet are equal to 1,445,674 meters?

Objective 2 Convert units of capacity.
For extra help, see Example 3 on pages 503–504 of your text and the Section 7.5 lecture video.
Convert.

5. A recipe calls for $\frac{3}{4}$ cup water. How many milliliters is this?

6. How many pints are there in 0.5 liters of milk?

7. A fuel tank holds 18.6 gallons of gasoline. How many liters is this?

8. A cough syrup says to take 15 milliliters. How many ounces is this?

Objective 3 Convert units of mass/weight.
For extra help, see Example 4 on page 504 of your text and the Section 7.5 lecture video.
Convert.

9. A box of cereal weighs 340 grams. Convert to ounces.

10. A pain reliever contains 400 milligrams of ibuprofen per does. How many ounces is this?

11. In some states a child must sit in a booster seat in a vehicle until the child weighs 40 pounds. How many kilograms is this?

12. It costs $0.39 to mail a letter weighing up to 1 ounce. How many grams is this?

Objective 4 Convert units of temperature.
For extra help, see Examples 5–6 on pages 505–506 of your text and the Section 7.5 lecture video.
Convert.

13. During January, the temperature is reported to be −19° F. What is this temperature in degrees Celsius?

14. A child has a temperature of 101.2° F. Convert this temperature to degrees Celsius.

15. To cook lasagna in an oven, preheat the oven to 350° F. What is this temperature in degrees Celsius?

16. On a sunny day in June the temperature is reported to be 22° C. Convert this temperature to degrees Fahrenheit.

Chapter 7 RATIOS, PROPORTIONS, AND MEASUREMENT

7.6 Applications and Problem Solving

Learning Objectives
1   Use debt-to-income ratios to decide loan qualification.
2   Calculate the maximum monthly PITI payment that meets the front-end ratio qualification for a loan.
3   Calculate the maximum debt that meets the back-end ratio qualification for a loan.
4   Calculate medical dosages.

Key Terms

Use the terms listed below to complete each statement in Exercises 1–2.

front-end   **back-end**   **interest**   **income**   **debt**   **total**

1. A PITI payment includes principle, _____, taxes, and insurance.

2. A(n) _____ ratio is the ratio of the total monthly debt payments to gross monthly income.

Use the following table for Guided Examples and Practice 1–3.

Factors for loan qualification	Conventional	VA	FHA
1. Front-end ratio should not exceed	0.28	N/A	0.29
2. Back-end ratio should not exceed	0.36	0.41	0.41
3. Credit score should be	650 or higher	N/A	N/A

GUIDED EXAMPLES AND PRACTICE

Objective 1   Use debt-to-income ratios to decide loan qualification.

Review this example for Objective 1:

1. The Haines family is trying to qualify for a conventional loan. Its gross monthly income is $4050. If the family gets the loan, the monthly payment will be $920. The family has monthly debt of $512 and a credit score of 727.
   a. What is the front-end ratio?
   b. What is the back-end ratio?
   c. Does the family qualify for the loan?

Practice this exercise:

1. The Shaw family is trying to qualify for a VA loan. Its gross monthly income is $4870. If the family gets the loan, the monthly payment will be $928. The family has monthly debt of $530 and a credit score of 703.
   a. What is the front-end ratio?

a.

$$\text{Front end ratio} = \frac{\text{monthly PITI payment}}{\text{gross monthly income}}$$

$$\text{Front end ratio} = \frac{920}{4050}$$

Front end ratio ≈ 0.227

Check:

$4050 \cdot 0.227 = 919.35 \approx 920$

b.

$$\text{Back-end ratio} = \frac{\text{total monthly debt payments}}{\text{gross monthly income}}$$

$$\text{Back-end ratio} = \frac{512 + 920}{4050}$$

Back-end ratio ≈ 0.3536

Check:

$4050 \cdot 0.3536 = 1432.08 \approx 512 + 920$

c. Since the front-end ratio and back-end ratio are both below conventional standards and the family has a great credit score, they should qualify for the loan.

b. What is the back-end ratio?

c. Does the family qualify for the loan?

Objective 2 Calculate the maximum monthly PITI payment that meets the front-end ratio qualification for a loan.

Review this example for Objective 2:

2. Suppose a family has a gross monthly income of $4300. Find the maximum mortgage payment that would meet the front-end ratio qualification for an FHA loan.

Practice this exercise:

2. Suppose a family has a gross monthly income of $4800. Find the maximum mortgage payment that would meet the front-end ratio qualification for a conventional loan.

$$\text{Front end ratio} = \frac{\text{monthly PITI payment}}{\text{gross monthly income}}$$

$$0.29 = \frac{p}{4300}$$

Replace with known values.

$$4300 \cdot 0.29 = \frac{p}{\cancel{4300}_{1}} \cdot \cancel{4300}^{1}$$

Multiply both sides by 4300.

$$1247 = p$$

Check:

$$\frac{1247}{4300} = 0.29$$

Objective 3 Calculate the maximum debt that meets the back-end ratio qualification for a loan.

Review this example for Objective 3:

3. Suppose a family has a gross monthly income of $4220. Find the maximum monthly debt that would meet the back-end ratio qualification for a conventional loan.

$$\text{Back-end ratio} = \frac{\text{total monthly debt payments}}{\text{gross monthly income}}$$

$$0.36 = \frac{d}{4220}$$

$$4220 \cdot 0.36 = \frac{d}{\cancel{4220}_{1}} \cdot \cancel{4220}^{1}$$

$$1519.20 = d$$

Check:

$$\frac{1519.20}{4220} = 0.36$$

Practice this exercise:

3. Suppose a family has a gross monthly income of $4310. Find the maximum monthly debt that would meet the back-end ratio qualification for an FHA loan.

Objective 4 Calculate medical dosages.

Review this example for Objective 4:

4. A patient with a mass of 57 kilograms is to receive an antibiotic. The order is to administer 8.9 milligrams per kilogram of mass. How much of the antibiotic should be administered?

Practice this exercise:

4. A patient is to receive 300 milliliters of a 5% D/W solution IV over 10 hours.

total dose = patient's mass · dosage per unit of mass

$$\text{total dose} = \frac{57 \ \cancel{kg}}{1} \cdot \frac{8.9 \ mg}{1 \ \cancel{kg}}$$

total dose = 507.3 mg

Check:

$$\frac{507.3 \ mg}{57 \ kg} = 8.9 \ mg / kg$$

The label on the box of the IV indicates that 10 drops dissipate 1 milliliter of the solution. How many drops should the patient receive each minute?

Use the following table for Additional Exercises 1–6.

Factors for loan qualification	Conventional	VA	FHA
1. Front-end ratio should not exceed	0.28	N/A	0.29
2. Back-end ratio should not exceed	0.36	0.41	0.41
3. Credit score should be	650 or higher	N/A	N/A

ADDITIONAL EXERCISES
Objective 1   Use debt-to-income ratios to decide loan qualification.
For extra help, see Example 1 on pages 510–511 of your text and the Section 7.6 lecture video.
Solve and check.

1.   The Martin family is trying to qualify for an FHA loan. Its gross monthly income is $3970. If the family gets the loan, the monthly payment will be $1012. The family has monthly debt of $726 and a credit score of 727.
     a.   What is the front-end ratio?
     b.   What is the back-end ratio?
     c.   Does the family qualify for the loan?

2.   The Holmes family is trying to qualify for a conventional loan. Its gross monthly income is $4100. If the family gets the loan, the monthly payment will be $933. The family has monthly debt of $508 and a credit score of 698.
     a.   What is the front-end ratio?
     b.   What is the back-end ratio?
     c.   Does the family qualify for the loan?

Objective 2 Calculate the maximum monthly PITI payment that meets the front-end ratio qualification for a loan.

For extra help, see Example 2 on page 511 of your text and the Section 7.6 lecture video.

Solve and check.

3. Suppose a family has a gross monthly income of $4600. Find the maximum mortgage payment that would meet the front-end ratio qualification for an FHA loan.

4. Suppose a family has a gross monthly income of $4200. Find the maximum mortgage payment that would meet the front-end ratio qualification for a conventional loan.

Objective 3 Calculate the maximum debt that meets the back-end ratio qualification for a loan.

For extra help, see Example 3 on page 512 of your text and the Section 7.6 lecture video.

Solve.

5. Suppose a family has a gross monthly income of $4260. Find the maximum monthly debt that would meet the back-end ratio qualification for a conventional loan.

6. Suppose a family has a gross monthly income of $4380. Find the maximum monthly debt that would meet the back-end ratio qualification for a VA loan.

Objective 4 Calculate medical dosages.

For extra help, see Examples 4–5 on pages 512–513 of your text and the Section 7.6 lecture video.

Solve.

7. A patient weighing 176 pounds is to receive a total of 0.5 milligrams per kilogram per day of accutane, given in two divided doses. If each tablet contains 10 milligrams, how many tablets should the patient receive in each dose?

8. A patient is to receive 600 milliliters of a 5% D/W solution IV over 5 hours. The label on the box of the IV indicates that 6 drops dissipate 1 milliliter of the solution. How many drops should the patient receive each minute?

Chapter 8 PERCENTS

8.1 Introduction to Percent

Learning Objectives
1 Write a percent as a fraction or decimal number.
2 Write a fraction or a decimal number as a percent.
3 Draw circle graphs.

Key Terms
Use the vocabulary terms listed below to complete each statement in Exercises 1–2.

left   **right**

1. When we convert from percent notation to decimal notation, we move the decimal point two places to the _____ .

2. When we convert from decimal notation to percent notation, we move the decimal point two places to the _____ .

GUIDED EXAMPLES AND PRACTICE
Objective 1   Write a percent as a fraction or decimal number.

Review these examples for Objective 1:
1. Write the percent as a fraction in lowest terms.

    15%

    $$15\% = \frac{15}{100} = \frac{3}{20}$$

Practice these exercises:
1. Write the percent as a fraction in lowest terms.

    40%

2. Write the percent as a decimal.

    16.2%

    $$16.2\% = \frac{16.2}{100} = 0.162$$

2. Write the percent as a decimal.

    95%

Objective 2   Write a fraction or decimal number as a percent.

Review these examples for Objective 2:

3. Write the fraction as a percent.

$$\frac{1}{4}$$

$$\frac{1}{4} = \frac{1}{\cancel{4}_1} \cdot \frac{\overset{25}{\cancel{100}}}{1} = 25\%$$

4. Write the decimal as a percent.

0.85

$$0.85 = (0.85)(100\%) = 85\%$$

Practice these exercises:

3. Write the fraction as a percent.

$$\frac{7}{9}$$

4. Write the decimal as a percent.

0.004

Objective 3   Draw circle graphs.

Review these examples for Objective 3:

5. Movie theater ticket sales show that 25% of the movies in the summer are comedies, 40% are action/adventure movies, and the rest are other kinds of movies. Draw a circle graph that represents this data.

Lightly draw 10 equal-size wedges so that each wedge represents 10%.

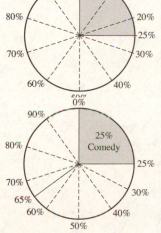

The wedge for 25% goes from 0% (straight up) to halfway between 20% and 30%.

The wedge for 40% is added to the wedge for 25%. 25% +40% = 65% Draw a radius line halfway between 60% and 70%.

Practice these exercises:

5. User ratings for a television go from five stars (the best) to one star (the worst.) On a website, the ratings for a TV were 20% 5-stars, 35% 4-stars, 15% 3-stars, 20% 2-stars and 10% 1-star. Draw a circle graph that represents these ratings.

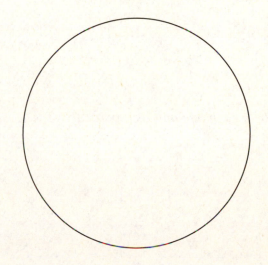

The rest of the graph represents other kinds of movies.

ADDITIONAL EXERCISES
Objective 1   Write a percent as a fraction or decimal number.
For extra help, see Examples 1–2 on pages 528–529 of your text and the Section 8.1 lecture video.
Write a percent as a fraction or decimal number.

1.   75% as a fraction.

2.   27.3% as a fraction

3.   4.8% as a decimal

4.   126% as a decimal.

Objective 2   Write a fraction or decimal number as a percent.
For extra help, see Examples 3–4 on pages 530–531 of your text and the Section 8.1 lecture video.
Write each fraction as a percent.

5. $\dfrac{14}{25}$

6. $\dfrac{7}{12}$

Write each decimal as a percent.

7.   0.83

8.   0.001

Name: Date:
Instructor: Section:

Objective 3   Draw circle graphs.
For extra help, see Example 5 on page 531 of your text and the Section 8.1 lecture video.
Draw the circle graph described.

9. The grades for a math course were
    15% As, 30% Bs, 35%Cs, 10% Ds
    and 10% Fs

10. A company found that 64% of its income
    went to salaries for the workers, 26% went
    to facilities, and the rest went to other
    expenses.

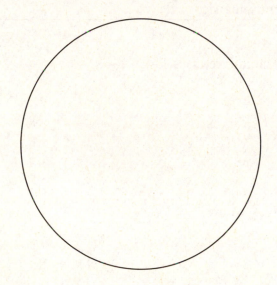

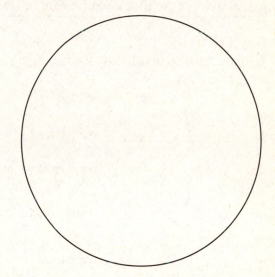

Chapter 8 PERCENTS

8.2 Solving Basic Percent Sentences

Learning Objectives
1   Identify the unknown amount in a basic percent sentence.
2   Solve for the unknown part in a basic percent sentence.
3   Solve for the unknown whole in a basic percent sentence.
4   Solve for the unknown percent in a basic percent sentence.

Key Terms
Use the terms listed below to complete each statement in Exercises 1–4.

%   is   of   what

1. _____ translates to "=."

2. _____ translates to any letter.

3. _____ translates to "$\times \dfrac{1}{100}$" or "$\times 0.01$."

4. _____ translates to "$\cdot$" or "$\times$."

GUIDED EXAMPLES AND PRACTICE

Objective 1   Identify the unknown amount in a basic percent sentence.

Review this example for Objective 1:
1. Identify the unknown amount as the percent, whole, or part.

   80% of 690 is what number?
   Answer: part

   A percent of   a whole is   a part.
   80%   of   690   is   what number?

Practice this exercise:
1. Identify the unknown amount as the percent, whole, or part.

   56% of what number is 25?

Objective 2   Solve for the unknown part in a basic percent sentence.

Review this example for Objective 2:
2.   30% of 72 is what number ?
     ↓   ↓   ↓   ↓   ↓
     0.3   · 72 = n
              21.6 = n

Practice these exercises:
2.   90% of 72 is what number?

Objective 3   Solve for the unknown whole in a basic percent sentence.

Review this example for Objective 3:	Practice this exercise:
3. 65% of what number is 130?	**3.** 36% of what number is 23.4?

For the review example:

3. 65% of what number is 130?
       ↓   ↓   ↓   ↓   ↓
      0.65 · n = 130

$$0.65n = 130$$

$$\frac{0.65n}{0.65} = \frac{130}{0.65}$$

$$n = 200$$

Objective 4   Solve for the unknown percent in a basic percent sentence.

Review this example for Objective 3:	Practice this exercise:
4. What percent of 350 is 70?	**4.** What percent of 40 is 16?

For the review example:

4. What percent of 350 is 70?
        ↓   ↓   ↓ ↓ ↓
       p · 350 = 70

$$350p = 70$$

$$\frac{350p}{350} = \frac{70}{350}$$

$$p = 0.2$$

$$p = (0.2)(100\%) = 20\%$$

ADDITIONAL EXERCISES

Objective 1   Identify the unknown amount in a basic percent sentence.

For extra help, see Example 1 on pages 534–535 of your text and the Section 8.2 lecture video.

Identify the unknown amount as the percent, whole, or part.

1. 18% of 93 is what number?

2. 5% of what number is 36?

3. What percent of 32 is 948?

4. What percent of 16 is 48?

Objective 2   Solve for the unknown part in a basic percent sentence.
For extra help, see Example 2 on pages 536–537 of your text and the Section 8.2 lecture video.
Translate to an equation and solve.

5. 45% of 120 is what number?

6. 135% of 16 is what number?

Objective 3   Solve for the unknown whole in a basic percent sentence.
For extra help, see Example 3 on page 538 of your text and the Section 8.2 lecture video.
Translate to an equation and solve.

7. 40% of what number is 62?

8. 64 is 96% of what number?

Objective 4   Solve for the unknown percent in a basic percent sentence.
For extra help, see Example 4 on pages 539–540 of your text and the Section 8.2 lecture video.
Translate to an equation and solve.

9. What percent of 81.5 is 32.6?

10. What percent of 21 is 14?

Chapter 8 PERCENTS

8.3 Solving Percent Problems (Portions)

Learning Objectives
1   Solve for the part in percent problems.
2   Solve for the whole amount in percent problems.
3   Calculate the percent.

GUIDED EXAMPLES AND PRACTICE

Objective 1   Solve for the part in percent problems.

Review this example for Objective 1:

1. Duchess earns 6% of total sales in commission. If she sells $1507 in merchandise during the week, what is her commission?

   The unknown is Duchess's commission, which is the part. We are given her commission rate (6%) and her total sales, ($1507) which is the whole.

   Her commission is 6% of her total sales.

   $$\downarrow \qquad \downarrow \ \downarrow \ \downarrow \ \downarrow$$
   $$c \qquad\quad = 0.06 \cdot 1506$$
   $$c = 90.36$$

   Duchess's commission is $90.36.

Practice this exercise:

1. Ten percent of Americans observe the "three-second rule" when eating food that has dropped to the floor. In other words, if it is picked up within three seconds, it is okay to eat it. If 220 Americans are asked if they observe the "three-second rule" how many are expected to observe the rule?

Objective 2 Solve for the whole amount in percent problems.

Review this example for Objective 2:

2. In a basketball game, the team made 40% of the total number of attempted shots. If they made a total of 40 shots, how many shots were attempted?

Practice this exercise:

2. *Communication Arts* magazine is sold to student subscribers for 68% of the regular subscription rate. If the student subscription price is $28.56, what is the regular subscription rate?

The unknown is the number of total
shots attempted. This is the whole.
We are given the percentage of shots
that the team made (40%) and the
number they made (40) which is the
part.

40% of how many shots is 40?

$\downarrow \quad \downarrow \qquad \downarrow \qquad \downarrow \ \downarrow$

$0.4 \quad \cdot \qquad s \qquad = \ 40$

$0.4s = 40$

$\dfrac{0.4s}{0.4} = \dfrac{40}{0.4}$

$s = 100$

The team attempted 100 shots.

Objective 3 Calculate the percent.

Review this example for Objective 3:

3. A company determines that 7 out of 125
    products are defective. What percent of the
    products are defective?

    The unknown is the percent. We are
    given the whole (125) and the part (7).

    What percent of 125 is 7?

    $\quad \downarrow \qquad\quad \downarrow \ \downarrow \ \downarrow \ \downarrow$

    $\quad p \qquad\quad \cdot \ 125 = \ 7$

    $\qquad\quad 125p = 7$

    $\dfrac{125p}{125} = \dfrac{7}{125}$

    $\qquad p = 0.056$

    $p = (0.056)(100\%) = 5.6\%$

    5.6% of the products are defective.

Practice this exercise:

3. Uber answers 32 out of 33
    questions on his test correctly.
    What percent of the questions
    did he answer correctly?

ADDITIONAL EXERCISES

Objective 1   Solve for the part in percent problems.

For extra help, see Examples 1–2 on pages 544–545 of your text and the Section 8.3 lecture video.

Translate to an equation and solve.

1. Amarillo, Texas had a total budget of $79 million during the 2003/2004 fiscal year. If 2.8% is spent on staff services, what amount is spent on staff services?

2. According to *USA Today*, girls make up 60% of the horror movie audience. If 425 people attend the showing of *Paranormal Activity 3*, how many are expected to be girls?

Objective 2   Solve for the whole amount in percent problems.

For extra help, see Example 3 on page 545 of your text and the Section 8.3 lecture video.

Translate to an equation and solve.

3. A dishwasher salesperson's commission rate is 30%. The salesperson receives a commission of $424. How many dollars worth of dishwashers did the salesperson sell?

4. Yesterday, 10 employees in an office were absent. If these absentees constitute 25% of the employees, how many employees are there?

Objective 3   Calculate the percent.

For extra help, see Example 4 on page 546 of your text and the Section 8.3 lecture video.

Translate to an equation and solve.

5. A blouse is normally sold for $42 and is discounted $14. What is the percent of discount?

6. The National Osteoporosis Foundation reports that 2,300,000 men and 7,800,000 women suffer from osteoporosis. What percent of the total number of people who suffer from osteoporosis are women?

Chapter 8 PERCENTS

8.4 Solving Problems Involving Percent of Increase or Decrease

---

Learning Objectives
1   Use a percent of increase to find an increase amount and the final amount after the increase.
2   Use a percent of decrease to find a decrease amount and the final amount after the decrease.
3   Find an unknown initial amount in a percent of increase or decrease problem.
4   Find a percent of increase or decrease.

Key Terms
Use the vocabulary terms listed below to complete Exercise 1. One of the terms will not be used.

   original amount   **new amount**   **amount of increase/decrease**

1.   We find a percent of increase/decrease by dividing the _____ by the
     _____.

GUIDED EXAMPLES AND PRACTICE

Objective 1   Use a percent of increase to find an increase amount and the final amount after the increase.

Review this example for Objective 1:	**Practice this exercise:**
1. The total dining bill is $42.60. If you decide to leave a 15% tip, how much money should you leave to cover the bill and the tip?	**1.** The sales tax rate in Massachusetts is 6.25%. If you bought a TV for $1138, find the sales tax and the total amount of the purchase.

First, find the amount of the tip.

  What is 15% of 42.60?

  $\downarrow$ $\downarrow$ $\downarrow$ $\downarrow$ $\downarrow$
  x $= 0.15$ $\cdot$ 42.6
   $x = 6.39$

The amount of the tip is $6.39.

The total amount needed for the bill and the tip is $42.60 + 6.39 = \$48.99$.

Copyright © 2012 Pearson Education, Inc. Publishing as Addison-Wesley

Objective 2   Use a percent of decrease to find a decrease amount and the final amount after the decrease.

Review this example for Objective 2:

2.   A student discount of 25% is applied to a purchase at a bookstore which had an initial price of $130. Find the amount of the discount and the final price.

First, find the amount of the discount.

What is 25% of 130?

$$\downarrow \quad \downarrow \quad \downarrow \quad \downarrow \quad \downarrow$$

$$x \quad = 0.25 \quad \cdot \; 130$$

$$x = 32.5$$

The amount of the discount is $32.50.

The final price after the discount is

$$130 - 32.50 = \$97.50$$

Practice this exercise:

2.   A sweater is on sale with a 30% discount. If the sweater has an initial price of $58.00, find the amount of the discount and the final price.

Objective 3   Find an unknown initial amount in a percent of increase or decrease problem.

Review this example for Objective 3:

3.   You added a 15% tip when charging a meal to a credit card. If the total amount charged to the card was $53.13, what was the cost of the meal before the tip?

The total bill = cost of the meal + tip amount

The total bill = cost of the meal + $0.15 \cdot$ cost of the meal

$$53.13 \quad = \quad c \quad \quad + 0.15c$$

$$53.13 = 1.15c$$

$$\frac{53.13}{1.15} = \frac{1.15c}{1.15}$$

$$c = 46.2$$

The cost of the meal was $46.20.

Practice this exercise:

3.   An employee's new salary is $16,800 after getting a 5% raise. What was the salary before the increase in pay?

Objective 4   Find a percent of increase or decrease.

Review this example for Objective 4:

4. The sales tax is $41 on the purchase of a dining room set originally priced at $820. Find the sales tax rate.

What percent of $820 is $41?
$$\downarrow \quad \downarrow \quad \downarrow \quad \downarrow \quad \downarrow$$
$$p \quad \cdot \quad 820 \quad = \quad 41$$

$$820p = 41$$

$$\frac{820p}{820} = \frac{41}{820}$$

$$p = 0.05$$

$$p = (0.05)(100\%) = 5\%$$

The sales tax rate is 5%.

5. A $385 suit is on sale for $231. Find the rate of discount.

$$\frac{\text{The rate of discount}}{100} = \frac{\text{discount amount}}{\text{initial amount}}$$

First, find the amount of the discount.

$$\$385 - \$231 = \$154$$

$$\frac{P}{100} = \frac{154}{385}$$

Equate the cross products to solve.

$$385P = 15400$$

$$\frac{385P}{385} = \frac{15400}{385}$$

$$p = 40$$

The rate of discount is 40%.

Practice this exercise:

4. Jackson earns $8.50 per hour. If his hourly pay is increased to $11.00, what is the percent of increase?

5. An item's original price is $700, but it is on sale for $140 less. Find the rate of discount and the sale price.

Name: Date:

Instructor: Section:

ADDITIONAL EXERCISES

Objective 1   Use a percent of increase to find an increase amount and the final amount after the increase.

For extra help, see Example 1 on pages 554–555 of your text and the Section 8.4 lecture video.

Solve.

1. The total dining bill is $23.48. If you decide to leave a 15% tip, how much money should you leave to cover the bill and the tip?

2. The sales tax rate in a city is 8.5%. Find the tax charged on a purchase of $227, and the total cost.

Objective 2   Use a percent of decrease to find a decrease amount and the final amount after the decrease.

For extra help, see Example 2 on page 556 of your text and the Section 8.4 lecture video.

Solve.

3. A $220 suit is on sale for a 10% discount. Find the amount of the discount and the sale price.

4. The state deducts 6% of the gross monthly pay for retirement. If Fitzgerald's gross monthly salary is $1586, how much is deducted for retirement? What is his pay after the deductions?

Objective 3   Find an unknown initial amount in a percent of increase or decrease problem.

For extra help, see Examples 3–4 on pages 557–558 of your text and the Section 8.4 lecture video.

Solve.

5. An item is marked down 20% from its original price. If this means that the item is selling for $21.40, what was the original price?

6. An amusement park offers a 15% military discount. If a soldier pays $55.25 for admission, what is the regular price?

Objective 4   Find a percent of increase or decrease.

For extra help, see Examples 5–6 on pages 558–560 of your text and the Section 8.4 lecture video.

Solve.

7. The amount in a savings account increased from $300 to $309. What was the percent of increase?

8. The population of a town decreased from 35,000 to 30,100. What was the percent decrease?

Chapter 8 PERCENTS

8.5 Solving Problems Involving Interest

Learning Objectives
1   Solve applied problems involving simple interest.
2   Solve applied problems involving compound interest.
3   Solve applied problems involving amortization.

Key Terms
Use the vocabulary terms listed below to complete each statement in Exercises 1–2.

compound interest   **simple interest**   **amortization**

1.   The formula $I = P \cdot r \cdot t$ is used to find the _____ I on principal P, invested for t years at interest rate r.

2.   When interest is paid on interest we call it _____.

3.   The process of paying off a loan in installments is called _____.

GUIDED EXAMPLES AND PRACTICE

Objective 1   Solve applied problems involving simple interest.

Review this example for Objective 1:

1.   What is the interest on $4400 invested at an interest rate of 8% for $\frac{1}{2}$ year?

Substitute $4400 for P, 8% for r, and $\frac{1}{2}$ for t in the simple interest formula.

$$I = P \cdot r \cdot t$$
$$= \$4400 \cdot 0.08 \cdot \frac{1}{2}$$
$$= \frac{\$4400 \cdot 0.08}{2}$$
$$= \$176$$

The interest for $\frac{1}{2}$ year is $176.

Practice this exercise:

1.   What is the interest on $1200 invested at 6% for $\frac{1}{4}$ year?

2. Kevin borrows $5000 for a car loan at 4% simple interest for 3 years. Find the amount of interest due and the total amount due after 3 years.

Find the amount of interest.
Substitute $5000 for P, 4% for r, and 3 for t in the simple interest formula.

$I = Prt$

$I = (5000)(0.04)(3)$

$I = 600$

Kevin will owe $600 in interest.

Find the total amount owed. Add the interest to the loan amount.
$5000 + \$600 = \$5600$

2. Sarah borrows $10,000 for a college loan at 6% simple interest for 5 years. Find the amount of interest due and the total amount due after 5 years.

Objective 2   Solve applied problems involving compound interest.

Review this example for Objective 2:

3. The Jensens invest $2000 in an account paying 12%, compounded semiannually. Find the amount in the account after $1\frac{1}{2}$ years.

Substitute $2000 for P, 12% for r, 2 for n, and $1\frac{1}{2}$ for t in the compound interest formula.

$$A = P \cdot \left(1 + \frac{r}{n}\right)^{n \cdot t}$$

$$= \$2000 \cdot \left(1 + \frac{0.12}{2}\right)^{2 \cdot \frac{3}{2}}$$

$$= \$2000 \cdot (1.06)^3$$

$$\approx \$2382.03$$

The amount in the account after $1\frac{1}{2}$ years is $2382.03.

Practice this exercise:

3. The Shaws invest $3600 in an account paying 8%, compounded quarterly. Find the amount in the account after 1 year.

Objective 3   Solve applied problems involving amortization.

Use the amortization table, figure 8.1 on page 572 of the text.

Review this examples for Objective 3:

4. Use Table 8.1 to find the monthly payment for $45,600 at 9% for 30 years.

    $45,600 does not appear on the table, so write it as a sum of values that are found in the table.

    $$45,000 + 500 + 100$$

    Now use the 30-year column to find the monthly payment for each of these principals. Add the monthly payments.

    $$45,000 + 500 + 100$$
    $$\downarrow \qquad \downarrow \qquad \downarrow$$
    $$362.08 + 4.02 + 0.80 = 366.90$$

    The monthly payment is $366.90.

Practice this exercise:

4. Use Table 8.1 to find the monthly payment on $375,700 at 9% for 15 years.

ADDITIONAL EXERCISES

Objective 1   Solve applied problems involving simple interest.

For extra help, see Examples 1–3 on pages 566–567 of your text and the Section 8.5 lecture video.

Find the simple interest.

	Principal	Rate of interest	Time	Simple interest
1.	$1200	3%	1 year	
2.	$500	4.8%	$\frac{1}{2}$ year	
3.	$25,000	$5\frac{3}{8}\%$	$\frac{1}{4}$ year	

Sheila borrows $2400 at 6% for 90 days.

4. Find the amount of interest due.

5. Find the total amount that must be paid after 90 days.

Objective 2   Solve applied problems involving compound interest.
For extra help, see Examples 4–5 on pages 569–570 of your text and the Section 8.5 lecture video.
Solve.

6. The Hanson's deposit $1000 in an account paying 5% compounded monthly. How much is in the account after 6 months?

7. Corey invests $3000 in an account paying 8%, compounded quarterly. How much is in the account after 2 years?

8. Lucas invests $10,000 in an account paying 6%, compounded quarterly. How much is in the account after 18 months?

9. Skylar deposits $5000 in an account paying 4% compounded daily. How much is in the account after 30 days?

Objective 3   Solve applied problems involving amortization.
Use the amortization table, Table 8.1 on page 572 of the text
For extra help, see Example 6 on pages 571 of your text, and Table 8.1 on page 572 of your text and the Section 8.5 lecture video.

10. Use Table 8.1 to find the monthly payment for $82,300 at 9% for 15 years.

11. Use Table 8.1 to find the monthly payment for $50,200 at 9% for 5 years.

12. Use Table 8.1 to find the monthly payment for $475,500 at 9% for 15 years.

13. Use Table 8.1 to find the monthly payment for $107,900 at 9% for 3 years.

Chapter 9 MORE WITH GEOMETRY AND GRAPHS

9.1 Points, Lines, and Angles

Learning Objectives
1   Identify points, lines, line segments, rays, planes, and parallel lines.
2   Identify angles.
3   Solve problems involving angles formed by intersecting lines.
4   Solve problems involving angles in a triangle.
5   Solve problems involving congruent triangles.

Key Terms
Use the terms listed below to complete each statement in Exercises 1–8.

supplementary	parallel	segment	point	exterior	plane
perpendicular	acute	right	obtuse	line	ray
complementary	vertical	angle	intersect	congruent	
corresponding	alternate	interior	straight	consecutive	

1. Lines that intersect to form a 90° angle are _____ lines.

2. A(n) _____ triangle has three angles that measure less than 90°.

3. A straight, one-dimensional figure extending forever in one direction from a single point is called a(n) _____.

4. _____ lines are in the same plane and never intersect.

5. A(n) _____ angle has measure 180°.

6. The sum of the measures of two _____ angles is 90°.

7. Vertical angles are _____ angles formed by the intersection of two lines.

8. In a figure formed by a transversal intersection two lines, a pair of angles between the lines on the same side of the transversal are called consecutive _____ angles.

Name: _____   Date: _____

Instructor: _____   Section: _____

GUIDED EXAMPLES AND PRACTICE

Objective 1 Identify points, lines, line segments, rays, planes, and parallel lines.

Review this example for Objective 1:
1. Identify the figure.

The two points, *M* and *N*, fix the line in place and are used to name the line.

line $\overleftrightarrow{MN}$

Practice this exercise:
1. Identify the figure.

Objective 2 Identify angles.

Review this example for Objective 2:
2. State whether each angle is acute, right, obtuse, or straight.

The angle is acute because its measure is between 0° and 90°.

Practice this exercise:
2. State whether each angle is acute, right, obtuse, or straight.

180°

Objective 3 Solve problems involving angles formed by intersecting lines.
For Exercise 3, use the following figure. Lines l and m are parallel.

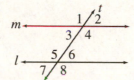

Review these examples for Objective 3:
3. Are ∠1 and ∠3 congruent or supplementary? Explain why.

Supplementary; because they form a straight angle.

Practice these exercises:
3. Are ∠2 and ∠3 congruent or supplementary? Explain why.

4. Four angles are formed by the intersecting lines. Find the measure of each angle.

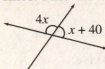

Because the two angles with variable expressions form a straight line, they are supplementary; so the sum of the measure is $180°$.

$4x + (x + 40) = 180$ Translate to an equation.

$5x + 40 = 180$ Combine like terms.

$\underline{-40 \quad -40}$ Subtract 40.

$5x + 0 = 140$

$5x = 140$

$\dfrac{5}{5} \quad \dfrac{}{5}$ Divide by 5.

$x = 28$

The first angle measures $4x = 4 \cdot 28 = 112°$. The second angle measures

$x + 40 = 28 + 40 = 68°$. Because the two intersecting lines form two pairs of vertical angles, the measures of the other angles are also $112°$ and $68°$.

Check.
The sum of the measures of supplementary angles is $112° + 68° = 180°$

4. Four angles are formed by the intersecting lines. Find the measure of each angle.

Objective 4 Solve problems involving angles in a triangle.

Review these examples for Objective 4:

5. Indicate whether the triangle is a right triangle, an acute triangle, or an obtuse triangle.

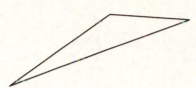

Obtuse triangle; the measure of one angle is greater than $90°$.

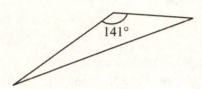

6. Find the measure of each angle in the following triangle.

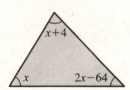

The angles of a triangle add up to $180°$.

$$x+(x+4)+(2x-64)=180$$
$$4x-60=180$$
$$4x=240$$
$$x=60°$$

Substitute $x=60$ to find the other two angles.

$$
\begin{array}{ll}
 & 2x-64 \\
x+4 & \\
60+4 \quad \text{and} & 2(60)-64 \\
 & 120-64 \\
64° & \\
 & 56°
\end{array}
$$

The three angles are $60°$, $64°$, and $56°$

Practice these exercises:

5. Indicate whether the triangle is a right triangle, an acute triangle, or an obtuse triangle.

6. Find the measure of each angle in the following triangle.

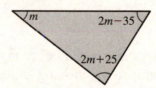

Name: Date:

Instructor: Section:

Objective 5   Solve problems involving congruent triangles.

Review these examples for Objective 5:	**Practice these exercises:**

7.   Determine whether the triangles are congruent. If they are congruent, state the rule that explains why.

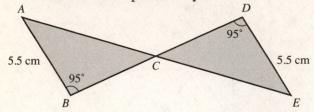

Sides $\overline{AB}$ and $\overline{DE}$ are congruent between the two triangles.

$\angle B$ and $\angle D$ are congruent.

The vertical angles at C are congruent,

The triangles are congruent by the AAS rule.

7.   Determine whether the triangles are congruent. If they are congruent, state the rule that explains why.

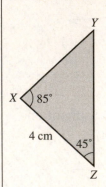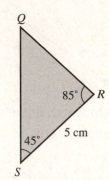

8.   The following triangles are congruent. Find the measure of the angles.

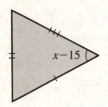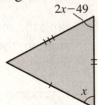

The given angle in the left triangle corresponds to the missing angle in the right triangle. The missing angle in the right triangle is $x-15$.

The angles of a triangle add up to $180°$.

$$x+(x-15)+(2x-49)=180$$
$$4x-64=180$$
$$4x=244$$
$$x=61°$$

Substitute $x=61$ to find the other two angles.

$x-15$ $2x-49$

$61-15$ and $2(61)-49$

$46°$ $122-49$

            $73°$

The three angles are $61°$, $46°$, and $73°$.

8.   The following triangles are congruent. Find the measure of the angles.

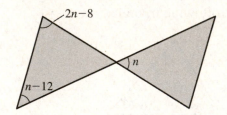

ADDITIONAL EXERCISES
Objective 1 Identify points, lines, line segments, rays, planes, and parallel lines.
For extra help, see Example 1 on page 585 of your text and the Section 9.1 lecture video.
Identify each figure.

1.

2.

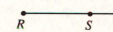

3.

4.

Objective 2 Identify angles.
For extra help, see Example 2 on pages 586–587 of your text and the Section 9.1 lecture video.
State whether each angle is acute, right, obtuse, or straight.

5.

6.

Objective 3 Solve problems involving angles formed by intersecting lines.
For extra help, see Examples 3–6 on pages 589–592 of your text and the Section 9.1 lecture video.
For exercises 7–10, use the following figure. Lines l and m are parallel.

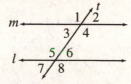

7. Are $\angle 4$ and $\angle 5$ congruent or supplementary? Explain why.

8. Are $\angle 4$ and $\angle 8$ congruent or supplementary? Explain why.

9. If $\angle 1 = 116°$, what is the measure of $\angle 8$?

10. If $\angle 6 = 52°$, what is the measure of $\angle 2$?

11. If $\angle 3 = 55°$, what is the measure of $\angle 6$?

12. If $\angle 5 = 121°$, what is the measure of $\angle 1$?

13. Lines h and k are parallel and the transversal t forms eight angles. Find the measure of each angle.

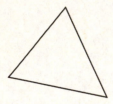

14. Lines l and m are parallel and the transversal t forms eight angles. Find the measure of each angle.

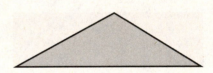

Objective 4 Solve problems involving angles in a triangle.
For extra help, see Examples 7–8 on pages 592–593 of your text and the Section 9.1 lecture video.

Indicate whether each triangle is a right triangle, an acute triangle, or an obtuse triangle. Explain why.

15.

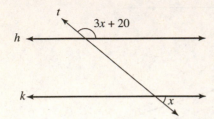

16.

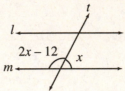

Find the measure of each angle in the following triangle.

17.

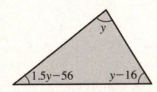

18.

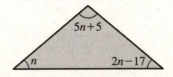

Objective 5 Solve problems involving congruent triangles.
For extra help, see Examples 9–10 on pages 595–596 of your text and the Section 9.1 lecture video.

Determine whether the triangles are congruent. If they are congruent, state the rule that explains why.

19.

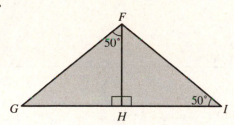

20.

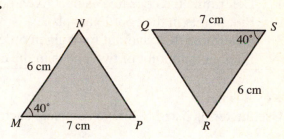

The following triangles are congruent. Find the measure of the angles.

21.

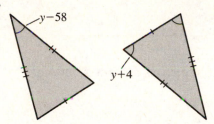

22.

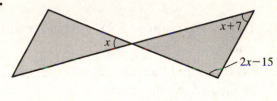

Chapter 9 MORE WITH GEOMETRY AND GRAPHS

9.2 The Rectangular Coordinate System

Learning Objectives
1   Determine the coordinates of a given point.
2   Plot points in the coordinate plane.
3   Determine the quadrant for a given ordered pair.
4   Find the midpoint of two points in the coordinate plane.

Key Terms
Use the terms listed below to complete each statement in Exercises 1–4.

> axis   quadrants   midpoint   coordinate   left   right

1.   In the coordinate plane, the intersection of the axes forms four _____.

2.   To plot the point $(-3, 9)$, first move three units to the _____.

3.   In an ordered pair, the second _____ indicates a point's vertical distance from the origin.

4.   A number line used to locate a point in the plane is called a(n) _____.

GUIDED EXAMPLES AND PRACTICE
Objective 1 Determine the coordinates of a given point.

Review this example for Objective 1:
1.   Determine the coordinates of each point shown.

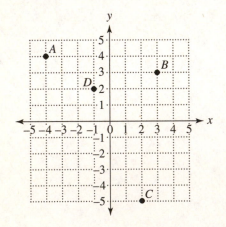

Practice this exercise:
1.   Determine the coordinates of each point shown.

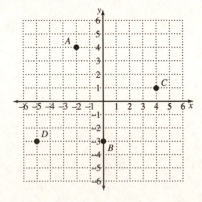

$A:(-4, 4)$ The vertical line through point A intersects the x-axis at -4. The horizontal line through point A intersects the y-axis at 4.

$B:(3, 3)$ The vertical line through point B intersects the x-axis at 3. The horizontal line through point B intersects the y-axis at 3.

$C:(2, -5)$ The vertical line through point C intersects the x-axis at 2. The horizontal line through point C intersects the y-axis at -5.

$D:(-1, 2)$ The vertical line through point D intersects the x-axis at -1. The horizontal line through point D intersects the y-axis at 2.

Objective 2   Plot points in the coordinate plane.

Review this example for Objective 2:

2.   Plot and label the point described by each ordered pair of coordinates.

$(0, 4)$, $(-1, -3)$, $(2, 6)$, $(-4, 3)$

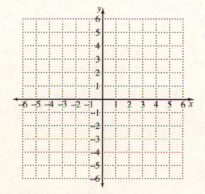

Practice this exercise:

2.   Plot and label the point described by each ordered pair of coordinates.

$(-4, -2)$, $(5, 0)$, $(2, -3)$, $(-5, 1)$

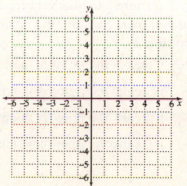

To plot $(0,4)$, begin at the origin, $(0,0)$, but don't move left or right since the first coordinate is 0, then move four units up.

To plot $(-1,-3)$, move from the origin one unit left because it is negative one, then three units down because it is negative three.

To plot $(2,6)$, move from the origin two units right, then six units up.

To plot $(-4,3)$, move from the origin four units left because it is negative four, then three units up.

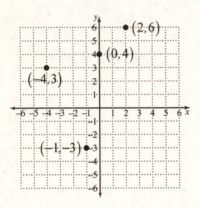

Objective 3 Determine the quadrant for a given ordered pair.

Review this example for Objective 3:

3. Determine the quadrant in which the point is located. $(-20,10)$

    To determine the quadrant for a given ordered pair, consider the signs of the coordinates.

    $(+,+)$ means the point is in quadrant I.

    $(-,+)$ means the point is in quadrant II.

    $(-,-)$ means the point is in quadrant III.

    $(+,-)$ means the point is in quadrant IV.

Practice this exercise:

3. Determine the quadrant in which the point is located. $(17,23)$

$(-20,10)$ is located in quadrant II (upper-left) because the first coordinate is negative and the second coordinate is positive.

Objective 4 Find the midpoint of two points in the coordinate plane.

Review this example for Objective 4:

4. Find the midpoint of the given points.

$(2,-7)$ and $(0,-2)$

Given two points with coordinates (x_1, y_1) and (x_2, y_2), the coordinates of their midpoint are $\left(\dfrac{x_1 + x_2}{2}, \dfrac{y_1 + y_2}{2} \right)$.

Let $(2,-7)$ be (x_1, y_1) and $(0,-2)$ be (x_2, y_2). Then the midpoint equals the following.

$= \left(\dfrac{2+0}{2}, \dfrac{-7+(-2)}{2} \right)$ Replace x_1, x_2, y_1, and y_2 with 2, 0, -7, and -2 respectively.

$= \left(\dfrac{2}{2}, \dfrac{-9}{2} \right)$ Add.

$= \left(1, -4\dfrac{1}{2} \right)$ Simplify.

Practice this exercise:

4. Find the midpoints of the given point.

$(3,-8)$ and $(7,10)$

Name: Date:
Instructor: Section:

ADDITIONAL EXERCISES
Objective 1 Determine the coordinates of a given point.
For extra help, see Example 1 on page 604 of your text and the Section 9.2 lecture video.
Determine the coordinates of each point shown.

1.

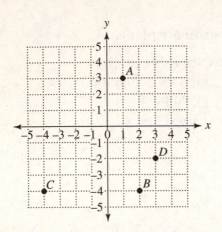

2.

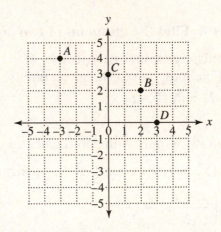

Objective 2 Plot points in the coordinate plane.
For extra help, see Example 2 on page 605 of your text and the Section 9.2 lecture video.
Plot and label the point described by each ordered pair of coordinates.

3. $(-2,4)$, $(-3,3)$, $(2,5)$, $(5,2)$

4. $(4,2)$, $(1,-5)$, $(-5,-3)$, $(-3,4)$

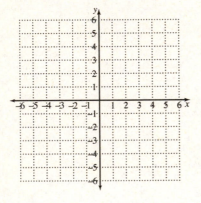

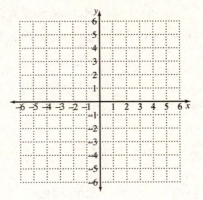

Objective 3 Determine the quadrant for a given ordered pair.

For extra help, see Example 3 on page 606 of your text and the Section 9.2 lecture video.

Determine the quadrant in which each point is located.

5. $(-33, -100)$ **6.** $(-18, 0)$

7. $(56, -117)$ **8.** $(0, 98)$

Objective 4 Find the midpoint of two points in the coordinate plane.

For extra help, see Example 4 on page 607 of your text and the Section 9.2 lecture video.

Find the midpoint of the given points

9. $(6, -10)$ and $(-2, -7)$ **10.** $(-10, 4)$ and $(-7, -4)$

11. $(-1, -10)$ and $(4, 10)$ **12.** $(7, 3)$ and $(-3, 5)$

13. $(-5.8, 2.8)$ and $(-2.1, 1.3)$ **14.** $(9, -7.2)$ and $(3.1, -7.9)$

Chapter 9 MORE WITH GEOMETRY AND GRAPHS

9.3 Graphing Linear Equations

Learning Objectives
1   Determine whether a given ordered pair is a solution for a given equation with two unknowns.
2   Find three solutions for an equation in two unknowns.
3   Graph linear equations, in x and y.
4   Graph horizontal and vertical lines.
5   Given an equation, find the coordinates of the x- and y-intercepts.

GUIDED EXAMPLES AND PRACTICE

Objective 1   Determine whether a given ordered pair is a solution for a given equation with two unknowns.

Review this example for Objective 1:

1. Determine whether the ordered pair is a solution for the equation.

    $(1,5); 2x + y = 7$

    $2(1) + (5) \overset{?}{=} 7$ Replace x with 1 and

    y with 5.

    $2 + 5 \overset{?}{=} 7$ Simplify.

    $7 = 7$ True.

    Because the resulting equation is true,
    $(1,5)$ is a solution for $2x + y = 7$.

Practice this exercise:

1. Determine whether the ordered pair is a solution for the equation.

    $(7,4); x - 2y = -1$

Objective 2   Find three solutions for an equation in two unknowns.
Objective 3 Graph linear equations, in x and y.

Review these examples for Objectives 2 and 3:

2. Find three solutions for the equation. Then graph. (Answers may vary for the three solutions.)

    $x - y = 3$

Practice these exercises:

2. Find three solutions for the equation. Then graph. (Answers may vary for the three solutions.)

    $3x - y = 6$

Let $x = 0$.   Let $x = 1$.   Let $x = 2$.
$0 - y = 3$ $1 - y = 3$ $2 - y = 3$
$-y = 3$ $-y = 2$ $-y = 1$
$y = -3$ $y = -2$ $y = -1$

List solutions to the equation in a table.

x	y	Ordered Pair
0	-3	$(0, -3)$
1	-2	$(1, -2)$
2	-1	$(2, -1)$

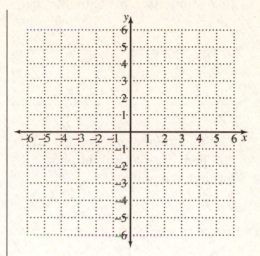

Plot the ordered pairs on the graph, and then draw a straight line through the points.

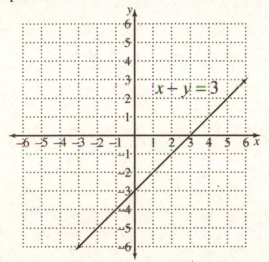

3. Find three solutions for the equation. Then graph. (Answers may vary for the three solutions.)
    $2x + 3y = 6$

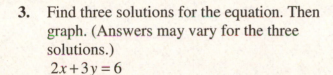

Let $x = 0$.   Let $x = 3$.   Let $x = 6$.
$2(0) + 3y = 6$ $2(3) + 3y = 6$ $2(6) + 3y = 6$
$0 + 3y = 6$ $6 + 3y = 6$ $12 + 3y = 6$
$3y = 6$ $3y = 0$ $3y = -6$
$y = 2$ $y = 0$ $y = -2$

3. Find three solutions for the equation. Then graph. (Answers may vary for the three solutions.)
    $3x + 4y = 12$

List solutions to the equation in a table.

x	y	Ordered Pair
0	2	$(0, 2)$
3	0	$(3, 0)$
6	-2	$(6, -2)$

Plot the ordered pairs on the graph, and then draw a straight line through the points.

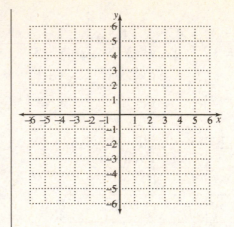

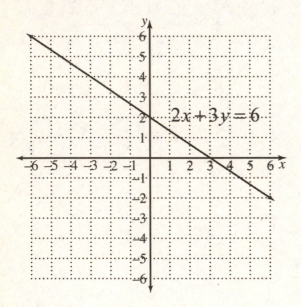

Objective 4 Graph horizontal and vertical lines.

Review this example for Objective 4:

4. Find three solutions for $y = 4$. Then graph.

All the solutions to this equation have a y-value of 4, no matter what values we pick for x.
Some possible solutions to the equation are $(-2, 4)$, $(0, 4)$, and $(1, 4)$.

Practice this exercise:

4. Find three solutions for $x = -3$. Then graph.

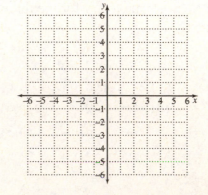

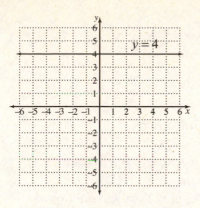

Objective 5 Given an equation, find the coordinates of the x- and y-intercepts.

Review this example for Objective 5:

5. Find the coordinates of the x- and y-intercepts.

$$4x + 6y = 48$$

For the x-intercept, replace y with 0 and solve for x.

$$4x + 6(0) = 48$$
$$4x + 0 = 48$$
$$4x = 48$$
$$x = 12$$

The x-intercept is $(12, 0)$.

For the y-intercept, replace x with 0 and solve for y.

$$4(0) + 6y = 48$$
$$0 + 6y = 48$$
$$6y = 48$$
$$y = 8$$

The y-intercept is $(0, 8)$.

Practice this exercise:

5. Find the coordinates of the x- and y-intercepts.

$$2x + 3y = 6$$

ADDITIONAL EXERCISES

Objective 1 Determine whether a given ordered pair is a solution for a given equation with two unknowns.

For extra help, see Examples 1–2 on pages 611–612 of your text and the Section 9.3 lecture video.

Determine whether the ordered pair is a solution for the equation.

1. $(-4, 2); y - 2x = 3$

2. $(-3, -2); y - x = 1$

3. $(0, -1); y = x$

4. $(0, 0); y = -\dfrac{3}{4}x$

Objective 2 Find three solutions for an equation in two unknowns.
Objective 3 Graph linear equations, in *x* and *y*.
For extra help, see Examples 3–7 on pages 613–615 of your text and the Section 9.3 lecture video.

Find three solutions for each equation. Then graph. (Answers may vary for the three solutions.)

5. $x + y = -5$

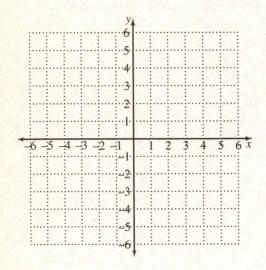

6. $2x - 3y = 8$

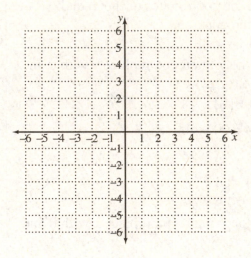

7. $y = -x$

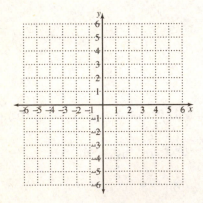

8. $y = -2x$

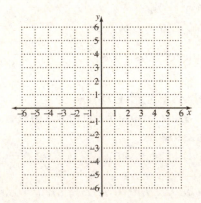

9. $y = 2x - 5$

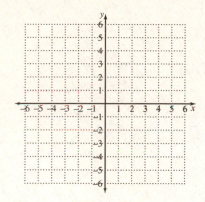

10. $y = -5x - 1$

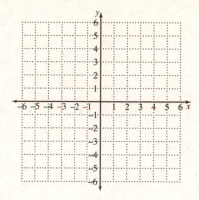

Objective 4 Graph horizontal and vertical lines.

For extra help, see Example 7 on pages 615–616 of your text and the Section 9.3 lecture video.

Find three solutions for each equation. Then graph. (Answers may vary for the three solutions.)

11. $y = -2$

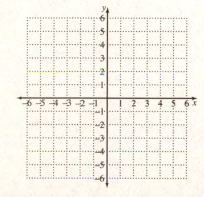

12. $x = 2$

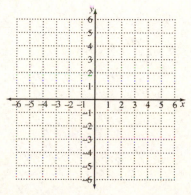

Objective 5 Given an equation, find the coordinates of the x- and y-intercepts.

For extra help, see Example 8 on page 617 of your text and the Section 9.3 lecture video.

Find the coordinates of the x- and y-intercepts.

13. $4x - 6y = 48$

14. $5x - 3y = 45$

15. $\dfrac{2}{5}x - 4y = 5$

16. $\dfrac{2}{3}x - 2y = 5$

17. $y = 9x$

18. $y = 1$

19. $x = -5$

20. $y = -4$

Chapter 9 MORE WITH GEOMETRY AND GRAPHS

9.4 Applications with Graphing

Learning Objectives
1   Solve problems involving linear equations in two variables.
2   Find the centroid of a figure given the coordinates of its vertices.
3   Find the area of a figure given the coordinates of its vertices.

Key Terms
Use the terms listed below to complete each statement in Exercises 1–2.

    vertex   centroid   balance   mean   coordinates

1.   The centroid of a figure is the _____ point of a figure.

2.   The coordinates of the centroid of a figure are given by the _____ of the corresponding coordinates.

GUIDED EXAMPLES AND PRACTICE
Objective 1 Solve problems involving linear equations in two variables.

Review this example for Objective 1:

1.   The linear equation
     $v = -32.8t + 570$ describes the velocity in feet per second of a rocket t seconds after being launched.

   a.   Find the initial velocity of the rocket.

   The initial velocity is the velocity at the time the rocket is launched, so when $t = 0$. Substitute 0 for t in the equation.

   $v = -32.8(0) + 570$

   $v = 0 + 570$

   $v = 570$

   The initial velocity is 570 feet per second.

Practice this exercise:

1.   The linear equation $v = -32.2t + 74$ describes the velocity in feet per second of a ball t seconds after being thrown straight up.

   a.   Find the initial velocity of the ball.

   b.   Find the velocity of the ball after 2.25 seconds.

   c.   How many seconds after launch will the ball stop before descending?

b. Find the velocity of the rocket after 8 seconds.

Substitute 8 for t in the equation.

$v = -32.8(8) + 570$

$v = -262.4 + 570$

$v = 307.6$

The velocity after 8 seconds is 307.6 feet per second.

c. How many seconds after launch will the rocket stop before returning to Earth?

The rocket will stop when $v = 0$.
Substitute 0 for v in the equation.

$0 = -32.8t + 570$

$-570 = -32.8t$

$17.38 = t$

The rocket will stop after 17.38 seconds.

d. Graph the equation with t as the horizontal axis and v as the vertical axis.

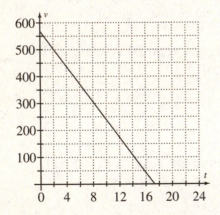

d. Graph the equation with t as the horizontal axis and v as the vertical axis.

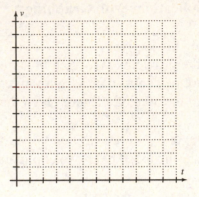

Objective 2   Find the centroid of a figure given the coordinates of its vertices.

Review this example for Objective 2:	**Practice this exercise:**
2. Find the centroid of the figure with the vertices. $(0,0)$, $(0,4)$, $(-3,4)$, $(-3,0)$	2. Find the centroid of the figure with the vertices. $(0,0)$, $(0,-8)$, $(-3,-8)$, $(-3,0)$

To find the centroid, find the mean (average) of all the x-coordinates of the vertices, and the mean of the y-coordinates of all the vertices.

$$\left(\frac{0+0+(-3)+(-3)}{4}, \frac{0+4+4+0}{4} \right)$$

$$\left(\frac{-6}{4}, \frac{8}{4} \right)$$

$$\left(-\frac{3}{2}, 2 \right)$$

The centroid is at $\left(-\dfrac{3}{2}, 2 \right)$.

Objective 3   Find the area of a figure given the coordinates of its vertices.

Review this example for Objective 3:	**Practice this exercise:**
3. Find the area of the figure with the vertices. $(0,0)$, $(2,7)$, $(9,7)$, $(7,0)$ Plot the points to find the shape of the figure.	3. Find the area of the figure with the vertices. $(0,0)$, $(4,8)$, $(9,8)$, $(5,0)$

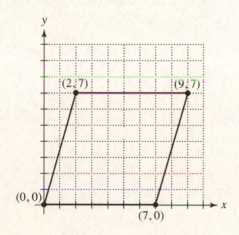

The figure is a parallelogram. Find the lengths needed to calculate the area.

The formula for the area of a parallelogram is $A = bh$ so we need to find the base and the height.

The base is $|0-7| = 7$

The height is $|7-0| = 7$

$A = 7 \cdot 7$

$A = 49$

The area is 49 square units.

ADDITIONAL EXERCISES
Objective 1 Solve problems involving linear equations in two variables.
For extra help, see Example 1 on pages 624–625 of your text and the Section 9.4 lecture video.
Solve.

1. The equation $b = 21.7t + 200$ describes the balance, in dollars, of an account t years after the initial investment is made.
   a. Find the initial balance.
   b. Find the balance after 8 years.
   c. Find the balance after 14 years.
   d. Graph the equation with t as the horizontal axis and b as the vertical axis.

2. The equation $b = 20.6t + 550$ describes the balance, in dollars, of an account t years after the initial investment is made.
   a. Find the initial balance.
   b. Find the balance after 9 years.
   c. Find the balance after 18 years.
   d. Graph the equation with t as the horizontal axis and b as the vertical axis.

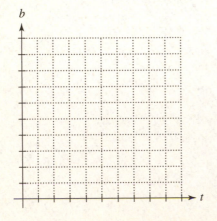

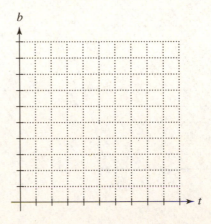

Objective 2 Find the centroid of a figure given the coordinates of its vertices.
For extra help, see Examples 2–3 on pages 626–627 of your text and the Section 9.4 lecture video.
Find the centroid of the figure with the given vertices.

3. $(-8,-8)$, $(-8,2)$, $(-4,2)$, $(-4,-8)$ **4.** $(-5,-1)$, $(-5,4)$, $(5,4)$, $(5,-1)$

Objective 3 Find the area of a figure given the coordinates of its vertices.
For extra help, see Example 4 on pages 627–628 of your text and the Section 9.4 lecture video.
Find the area of the figure with the given vertices.

5. $(0,0)$, $(0,6)$, $(4,6)$, $(7,0)$ **6.** $(0,0)$, $(0,8)$, $(5,8)$, $(7,0)$

Odd Answers to MyWorkBook

Chapter 1

Section 1.1

Key Terms

1. whole numbers
2. set
3. statistic
4. natural numbers
5. inequality
6. subset
7. equation

Practice

1. 8
2. billions
3. 3 ten millions + 5 millions + 6 hundred thousands + 9 thousands + 9 hundreds + 7 tens + 2 ones
4. 90,238
5. four hundred sixteen thousand, nine hundred ninety-five
6.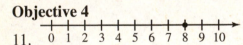
7. <
8. 456,583,100
9. 300 votes

Objective 1

1. 0
3. ones

Objective 2

5. 1 ten thousands + 4 thousands + 3 hundreds + 2 tens + 1 ones
7. 90,238

Objective 3

9. three hundred fifteen thousand, four hundred ninety

Objective 4

11.
```
 +--+--+--+--+--+--+--+--●--+--+-->
 0  1  2  3  4  5  6  7  8  9  10
```

Objective 5

13. =

Objective 6

15. 19,000,000

Objective 7

17. $6730; $6145

Section 1.2

Key Terms

1. solution
2. perimeter
3. variable
4. related equation
5. subtraction
6. addition
7. constant

Practice
1. 71,439
3. $27,150
5. 455
7. 1567 people

2. estimate: 90,000; actual: 89,778
4. 50 ft.
6. $r = 13$
8. 30 GB

Objective 1
1. 483

3. 96

Objective 2
5. estimate: 112,000; actual: 112,079

Objective 3
7. $121

Objective 4
9. 5377

Objective 5
11. $t = 56$

Objective 6
13. 118 squares

Objective 7
15. $10,559

Section 1.3

Key Terms
1. area
3. rectangular array
5. multiplication
7. formula
9. square unit

2. exponential form
4. base
6. distributive property
8. exponent

Practice
1. 50
3. 1060 mi.
5. 3^{11}
6. $6\times10^7 + 2\times10^6 + 8\times10^5 + 1\times10^3 + 4\times10 + 3\times1$
7. 17,576,000 tags

2. estimate: 16,000; actual: 15,516
4. 10,000

Objective 1
1. 62,624

Objective 2
3. estimate: 320,000; actual: 358,224

Objective 3
5. 840 cupcakes

Objective 4
7. 1

9. 343

Objective 5

11. 21^4 13. 15^3

15. $7 \times 10^5 + 8 \times 10^4 + 5 \times 10^3 + 4 \times 10^2 + 6 \times 10 + 3 \times 1$

Objective 6

17. 378

Section 1.4

Key Terms

1. divisor
3. dividend

2. division
4. remainder

Practice

1. 0; because $8 \cdot 0 = 0$
3. yes
5. $x = 11$
6. 70 bags can be filled; 17 kg will be left over.

2. no
4. 986 r2

Objective 1

1. 27 because $27 \times 1 = 27$
5. no
9. 146 r4

3. no
7. yes
11. 320

Objective 2

13. $t = 0$

15. $x = 9$

Objective 3

17. 4 feet

Section 1.5

Key Terms

1. mean (or average)
3. mode

2. average
4. median

Practice

1. 37
3. mean: 54; median 54; mode: 54

2. 24

Objective 1

1. 2
5. 76
9. 3060

3. 640
7. 46

Objective 2
11. mean: 79; median: 79; mode: 78 and 80
13. mean: 555; median: 562; mode: none

Section 1.6

Key Terms
1. right angle
2. volume
3. parallel lines
4. cubic unit
5. parallelogram

Practice
1. 48 in.
2. 1040 m^2
3. 450 in.^{2}
4. 12 ft.
5. 2744 ft.^{2}

Objective 1
1. 624 ft.

Objective 2
3. 465 cm^2

Objective 3
5. 910 cm^3

Objective 4
7. 21 cm
9. 2 cm

Objective 5
11a. 499 ft.^{2} b. 6 gal.   c. $138

Chapter 2

Section 2.1

Key Terms
1. absolute value
2. negative
3. integers
4. additive inverse
5. positive

Practice
1. −21
2. +1350
3.
4. 18° F
5. <
6. 59
7. 33
8. 46
9. 17

Objective 1

1. +416

3. −1

Objective 2

5. 0 1 2 3 4 5 6 7 8 9 10

7. −10 −9 −8 −7 −6 −5 −4 −3 −2 −1 0

Objective 3

9. <

11. >

Objective 4

13. 26

15. 175

Section 2.2

Practice

1. −32

2. −10

3. A debt with a credit, so subtract, and because the debt, −63, is more than the credit, the result is a debt/negative; −22.

4. 11

5. 83

6. Their new altitude is −92 m.

Objective 1

1. 27

3. −11

5. −39

Objective 2

7. −9

9. −12

Objective 3

11. −12

13. 4

Objective 4

15. 74

17. −4

Objective 5

19. $358

21. 65 feet

Section 2.3

Key Terms

1. cost

2. net

3. loss

4. revenue

Practice

1. $15 + (-37) = -22$

2. $-14 + 16 = 2$

3. $x = -16$

4. $d = 30$

5. The submarine ascended 1043 ft.

Objective 1

1. $8 + (-4) = -4$

3. $21 + (-16) = 5$

Objective 2

5. 12

7. 83

Objective 3

9. Profit of $1,000,500

Section 2.4

Key Terms

1. even

2. negative

3. positive

4. same

5. positive

6. negative

Practice

1. -40

2. 50

3. -15

4. -8

5. -25

6. 7

7. $f = 6$

8. ± 13

9. -11

10. Kevin pays a total of $40,500 per year in tuition.

Objective 1

1. -32

3. 0

5. -240

Objective 2

7. 16

9. -25

Objective 3

11. -9

13. undefined

Objective 4

15. $y = 3$

17. $r = -18$

Objective 5

19. 12 and -12

21. 5

Objective 6
23. $136

Section 2.5

Practice
1. −14
3. 149
5. 4

2. 5
4. −12

Objective 1
1. −12
5. −66
9. 4
13. −19

3. −2
7. 33
11. 12
15. 0

Section 2.6

Practice
1. $5006; profit
3. The plane travels 453 km.

2. −495 V

Objective 1
1. profit of $125

Objective 2
3. −64 amps

Objective 3
5. 52 mph

Chapter 3

Section 3.1

Key Terms
1. evaluate
3. indeterminate
5. undefined

2. equation
4. expression

Practice
1. equation

3. $5n^2$

5. 18

2. expression

4. $\dfrac{x-y}{x+4}$

6. 1

Objective 1
1.   Equation

Objective 2
3. $-3(n-4)$

Objective 3
5.   0
9. -15
13. -27
15.

x	$-3x+5$
-2	11
-1	8
0	5
1	2
2	-1

7.   33
11. 4

17. $x=4$

Section 3.2

Key Terms
1.   greatest
3.   coefficient
5.   degree

2.   simplest
4.   trinomial
6.   monomial

Practice
1.   not a monomial; not the product of a constant and a variable raised to a whole number power
2.   coefficient: -1; degree: 1
3.   first term: $3c^5$; coefficient: 3, second term: $-8c^3$, coefficient: -8; third term: -18, coefficient: -18
4.   no special name
5.   degree: 5
6. $14z^3+12z^2+z-15$
7.   not like
8. $-a$
9. $-2c^4+3c^2-2$

Objective 1
1.   not a monomial; not the product of a constant and a variable raised to a whole number power

Objective 2
3. coefficient: 7, degree: $4+2=6$

Objective 3
5. first term: $4y^5$; coefficient: 4, second term: $13y^4$, coefficient: 13; third term: $-y^3$, coefficient: -1; fourth term: -12, coefficient: -12
7. trinomial

Objective 4

9. degree: 3

Objective 5

11. $-z^6 - 4z^5 + 3z^2 - z + 15$

Objective 6

13. not like

17. $10p$

21. $3x^5 + 2x + 8$

25. $11h^3 - 22h^2 - 4h$

15. $-4y^4$

19. $13x$

23. $-8v^4 + 8v^2 + v + 25$

Section 3.3

Practice

1. $12k^2 - k - 8$

2. $17y^2 + 9y + 7$

Objective 1

1. $11y - 11$

5. $14u^5 + 2u^3 + 4u^2 - 3$

3. $13x^3 + 6x + 17$

7. $4x + 34$

Objective 2

9. $x - 1$

13. $-2u^4 + 15u^2 - 15u - 17$

11. $35u^3 - 29u^2 + 17u + 5$

Section 3.4

Key Terms

1. multiply

3. multiply; add

2. exponents; base

4. sign

Practice

1. x^{17}

3. $-28x + 35$

5. $2y^3 + 4y^2 - 27y + 15$

2. $4096k^{42}$

4. $4b^2 - 9$

Objective 1

1. p^3

3. $24y^5$

Objective 2

5. $-27w^{15}$

Objective 3

7. $18y - 6$

9. $-8a^7 + 12n^6 - 4n^5$

Objective 4

11. $u^2 + u - 12$

13. $10x^2 - 17xy + 3y^2$

15. $6s^2 + st - 77t^2$

17. $a^2 - 49$

19. $-18b^4 - 24b^3 + 24b^2$

21. $21y^2 + 12y$

Section 3.5

Key Terms

1. prime

2. prime

3. greatest

4. composite

Practice

1. not prime

2. prime

3. $200 = 2^3 \cdot 5^2$

4. 1, 2, 3, 6, 9, 18, 27, 54

5. GCF $= 1$

6. GCF $= 1$

7. $10x^5$

Objective 1

1. prime

3. not prime

Objective 2

5. $512 = 2^9$

7. $2205 = 3^2 \cdot 5 \cdot 7^2$

Objective 3

9. 1, 3, 5, 15, 25, 75, 125, 375

Objective 4

11. GCF $= 1$

Objective 5

13. GCF $= 20$

Objective 6

15. $6x^3$

Section 3.6

Key Terms

1. product

2. divisor's; dividend's

3. quotient

Practice

1. $3x^2$

2. $6x^7 - 8x^5 + 5$

3. $7a$

4. $25x + 3x^2$

5. $2(c + 16)$

Objective 1

1. $-9x^2$

3. $4y^4$

Objective 2

5. $6x + 3$

7. $-9x^7 + 5x^2 - 5$

Objective 3

9. $8y$

11. $-4d^5 + 5d$

13. $9y$

Objective 4

15. $5(a - 3)$

17. $b(2 - 7b^3)$

19. $4x(4x^2 - 3x + 1)$

21. $x^3(7x^3 - 19x + 5)$

Section 3.7

Practice

1. $24x + 2$

2. 54

3. 576

4. a. $N = 33x + 6y - 10$; b. $15{,}566$

Objective 1

1. $20x + 8$; 68 yards

3. $12x^2 - 25x - 7$; 561

5. $x^3 + 10x^2 + 21x$; 90

Objective 2

7. 324 cm^2

Objective 3

9. After 25 seconds the skydiver has an altitude of 5000 feet.

Objective 4

11. a. $10s + 12m + 16l$; b. $4s + 6m + 7l$; c. $6s + 6m + 9l$; d. $\$228$

Chapter 4

Section 4.1

Key Terms

1. equation

2. solve

3. expression

4. solution

Practice

1. equation

2. expression

3. 21 is a solution.

4. 2 is not a solution.

Objective 1

1. equation

3. expression

Objective 2

5. -6 is not a solution.

7. 7 is a solution.

9. 10 is not a solution.

11. 9 is not a solution.

13. 8 is not a solution.

15. 8 is a solution.

17. 9 is not a solution.

Section 4.2

Key Terms

1. solution

2. nonlinear

3. add

4. distributive

Practice

1. linear

2. $y = -20$

3. $h = 9$

4. $346 = 168 + 148 + x$; $30

Objective 1

1. linear

Objective 2

3. $n = 12$

5. $t = 6$

7. $w = 30$

Objective 3

9. $35 + 33 + l = 86$; 18 mm

Section 4.3

Practice

1. $y = 2$

2. $m = -1$

3. $c = -2$

4. $v = -5$

5. The travel time will be 5 hr.

Objective 1

1. $a = -5$

3. $r = 3$

Objective 2

5. $s = 2$

7. $r = -8$

9. $q = 2$

11. $p = -2$

13. $x = 2$

Objective 3

15. The width is 20 cm.

Section 4.4

Practice
1. $n-7=14$; $n=21$
2. $-2y-14=16$; $y=-15$
3. $6n+4=4(n+12)$; $n=22$

Objective 1
1. $n-14=10$; $n=24$
3. $n+11=-19$; $n=-30$
5. $-5n=35$; $n=-7$
7. $8n=-16$; $n=-2$
9. $4t+11=6t-13$; $t=12$
11. $-10(y-15)-20=-4y-(y-5)$; $y=25$
13. $7(m+3)-6m=3(m+3)$; $m=6$

Section 4.5

Key Terms
1. complementary
2. supplementary
3. isosceles
4. congruent

Practice
1. $n+6n=266$; the numbers are 38 and 228.
2. $l+l+l=l+(4l-35)+l+(4l-35)$; each side of the equilateral triangle is 10 ft long; the rectangle is 5 ft wide and 10 ft long.
3. $a+a+a+39=180$; angle A is $47°$, angle B is $47°$, and angle C is $86°$.
4. $12(20B)+13B=1265$; the price of stock A is \$100 and the price of stock B is \$5.

Objective 1
1. $(a+5)+a=29$; the numbers are 17 and 12.
3. $l+l+l=l+(3l-30)+l+(3l-30)$; each side of the equilateral triangle is 12 ft long, the rectangle is 6 ft wide and 12 ft long.

Objective 2
5. $5s+3(s+6)=170$; each knit shirt costs \$19 and each pair of pants costs \$25.
7. $9c+6(c+75)=1725$; cats cost \$85 and dogs cost \$160.

Chapter 5

Section 5.1

Key Terms
1. fewest
2. numerator
3. multiple
4. rational
5. mixed

Practice

1. $\dfrac{7}{10}$

2. $\dfrac{31}{61}$

3. 0

4. $5\dfrac{3}{4}$

5. $-8\dfrac{5}{7}$

6. $-\dfrac{32}{3}$

7.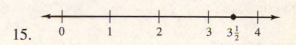

8.

9. -28

10. 3

11. $<$

Objective 1

1. $\dfrac{1}{2}$

3. $\dfrac{1}{12}$

Objective 2

5. -17

7. $7\dfrac{5}{6}$

9. $12\dfrac{11}{14}$

Objective 3

11. $-\dfrac{17}{7}$

Objective 4

13.

15.

Objective 5

17. 40

Objective 6

19. <

21. >

Section 5.2

Key Terms

1. rational

2. lowest

3. greatest

4. numerator; denominator

Practice

1. $\dfrac{3}{5}$

2. $-\dfrac{4}{5}$

3. $\dfrac{5}{8}$

4. $3\dfrac{1}{3}$

5. $4\dfrac{3}{4}$

6. $\dfrac{2k}{3}$

7. $\dfrac{3b^2 d^3}{5c^3}$

Objective 1

1. $\dfrac{3}{5}$

3. $-\dfrac{7}{9}$

5. $\dfrac{5}{9}$

7. $\dfrac{17}{20}$

9. $\dfrac{10}{11}$

Objective 2

11. $1\dfrac{7}{11}$

13. $-1\dfrac{5}{13}$

Objective 3

15. $\dfrac{3y}{4}$

17. $-\dfrac{3c}{4d}$

19. $\dfrac{1}{35n}$

Section 5.3

Key Terms

1. irrational

2. circle

3. circumference

4. diameter

Practice

1. $\dfrac{9}{28}$

2. $-\dfrac{8}{35}$

3. $\dfrac{21}{40}$

4. $\dfrac{1}{12}$

5. $\dfrac{2}{11}$

6. $10;\ 9\dfrac{3}{4}$

7. $5;\ 6\dfrac{1}{4}$

8. $11\dfrac{1}{2}$

9. $\dfrac{5}{12x^2}$

10. $\dfrac{16}{6561}$

11. $\dfrac{81h^8}{64\,j^4 k^2}$

12. 217 votes

13. $\dfrac{3}{8}$

14. $9\dfrac{3}{8}$ in.^{2}

15. $9\dfrac{2}{3}$ in.

16. $1\dfrac{4}{7}$ m

17. $16\dfrac{1}{2}$ ft

Objective 1

1. $-\dfrac{1}{48}$

3. $\dfrac{9}{1000}$

Objective 2

5. $\dfrac{10}{21}$

7. $-\dfrac{5}{52}$

9. $\dfrac{1}{6}$

Objective 3

11. estimate: 18; actual: $19\dfrac{11}{20}$

13. $3\dfrac{55}{64}$

Objective 4

15. $\dfrac{3z^2}{20}$

17. $-\dfrac{3y^2}{x^3 z^2}$

Objective 5

19. $\dfrac{25}{16}$

21. $\dfrac{w^{36}}{256}$

Objective 6

23. Cheryl will earn \$12 for working $\dfrac{3}{8}$ of a day.

Objective 7

25. 324 mm^2

Objective 8

27. $1\dfrac{1}{4}$ in.

Objective 9

29. $4\dfrac{2}{5}$ in.

Section 5.4

Key Terms

1. complex
3. denominator

2. reciprocal
4. improper

Practice

1. $\dfrac{1}{4}$

2. $-\dfrac{3}{22}$

3. $14;\ 16\dfrac{1}{2}$

4. $\dfrac{8r^3}{3pq^6}$

5. $\dfrac{4}{9}$

6. 4

7. $y = 2\dfrac{4}{5}$

8. $k = 31\dfrac{1}{3}$

9. $1\dfrac{5}{8}$

Objective 1

1. $\dfrac{9}{4}$

3. $-\dfrac{1}{11}$

5. 3

7. $-\dfrac{3}{13}$

9. $5\dfrac{7}{9}$

Objective 2

11. estimate: $\dfrac{3}{2}$; actual: $1\dfrac{11}{34}$

13. estimate: 2; actual: $2\dfrac{31}{44}$

Objective 3

15. $\dfrac{2x^3}{5y^4}$

Objective 4

17. $\dfrac{5}{6}$

19. 3

Objective 5

21. 28

Objective 6

23. 10 applications can be made.

Section 5.5

Key Terms

1. multiple

2. least

3. greatest

Practice

1. 105

2. 5040

3. $216b^2c^3d^2$

4. $\dfrac{25}{30}; \dfrac{26}{30}$

5. $\dfrac{22}{77m^5}; \dfrac{42m^2}{77m^5}$

Objective 1

1. 60

3. 225

Objective 2

5. 36
9. 378

7. 441

Objective 3

11. $225xy$

13. $147x^9z^4$

Objective 4

15. $\dfrac{66}{126}$ and $\dfrac{49}{126}$

17. $\dfrac{39}{45}$ and $\dfrac{41}{45}$

Objective 5

19. $\dfrac{10y}{45xy}$ and $\dfrac{33x}{45xy}$

21. $\dfrac{282y^6}{441x^6y^8}$ and $\dfrac{287x^2}{441x^6y^8}$

Section 5.6

Practice

1. $\dfrac{1}{4}$

2. $\dfrac{4}{9}$

3. $\dfrac{6s}{7}$

4. $\dfrac{7b-3}{10}$

5. $\dfrac{11}{30}$

6. $\dfrac{12+15w}{40w}$

7. $\dfrac{33z^2-14}{60z}$

8. $9;\ 8\dfrac{11}{12}$

9. $3;\ 2\dfrac{23}{24}$

10. $2\dfrac{3}{8}$

11. $f = \dfrac{17}{24}$

12. $5\dfrac{29}{30}$ miles per gallon

Objective 1

1. $\dfrac{4}{5}$

2. $\dfrac{13}{15}$

3. $-\dfrac{11}{14}$

4. $\dfrac{13}{20}$

Objective 2

5. $\dfrac{8}{y}$

6. $\dfrac{18}{mn}$

7. $\dfrac{7p}{16}$

8. $\dfrac{5q}{h^2}$

9. $\dfrac{4-n}{21m}$

10. $\dfrac{3m+5n}{8k}$

Objective 3

11. $\dfrac{7}{33}$

12. $\dfrac{13}{36}$

13. $-\dfrac{1}{8}$

14. $\dfrac{149}{120}$

Objective 4

15. $\dfrac{23}{14x}$

16. $\dfrac{65q+88}{110q}$

17. $\dfrac{12-2y}{21y}$

18. $\dfrac{9b^2-5}{12b^3}$

Objective 5

19. $14\dfrac{2}{3}$

20. $5\dfrac{5}{7}$

21. $6\dfrac{3}{4}$

22. $3\dfrac{47}{55}$

23. $6\dfrac{11}{18}$

24. $-3\dfrac{1}{4}$

25. $-5\dfrac{1}{30}$

26. $-13\dfrac{9}{10}$

Objective 6

27. $3\dfrac{11}{21}$

28. $-\dfrac{11}{15}$

29. $-15\dfrac{13}{18}$

30. $3\dfrac{1}{15}$

Objective 7

31. $y=-\dfrac{1}{4}$

32. $m=-\dfrac{37}{63}$

33. $q=\dfrac{53}{44}$

34. $p=\dfrac{15}{26}$

Objective 8

35. The new tire tread is $\dfrac{1}{32}$ inch deeper.

36. The fourth child inherited $\dfrac{1}{9}$ of the estate.

37. The distance around the book is $31\dfrac{3}{25}$ inches.

38. Phil would have biked $1\dfrac{13}{60}$ fewer miles.

Section 5.7

Practice

1. $\dfrac{3}{4}$

2. $-\dfrac{1}{5}$

3a. $6\dfrac{1}{6}$ minutes

3b. $6\dfrac{1}{6}$ minutes

3c. $5\dfrac{1}{3}$ minutes

4. $3\dfrac{7}{9}$

5. $70\dfrac{7}{8}$ ft^2

6. $1\dfrac{43}{56}$ m^2

7. $\dfrac{13}{14}j^3 - \dfrac{7}{10}j^2 + \dfrac{3}{4}$

8. $\dfrac{11}{40}u^2 + \dfrac{29}{63}v$

9. $\dfrac{1}{24}n^5$

10. $\dfrac{5}{12}s^2 + \dfrac{3}{2}s - 24$

Objective 1

1. $-4\dfrac{1}{10}$

3. $2\dfrac{7}{50}$

5. $-11\dfrac{3}{4}$

Objective 2

7a. height: $46\dfrac{5}{8}$ in.; weight: $49\dfrac{13}{24}$ lbs.

7b. height: $46\dfrac{7}{8}$ in.; weight: $49\dfrac{7}{8}$ lbs.

7c. height: $47\dfrac{1}{2}$ in.; weight: $42\dfrac{1}{4}$ lbs.

Objective 3

9. $6\dfrac{1}{4}$

11. $-1\dfrac{17}{48}$

Objective 4

13. 7 m^2

Objective 5

15. $254\dfrac{4}{7}$ cm^2

Objective 6

17. $-\dfrac{11}{12}y^3 + \dfrac{5}{8}y + \dfrac{27}{4}$

19. $\dfrac{11}{12}n^2 - \dfrac{4}{5}n + \dfrac{19}{12}$

21. $-\dfrac{5}{14}m^8$

23. $\dfrac{5}{2}n - \dfrac{10}{21}$

25. $-\dfrac{1}{32}x^5 + \dfrac{9}{4}x^4 - \dfrac{1}{24}x^3$

27. $2x^2 - 11x - 6$

Section 5.8

Practice

1. $k = \dfrac{7}{8}$

2. $3\dfrac{1}{3}n + \dfrac{3}{8} = \dfrac{7}{9}n - 1\dfrac{1}{6};\ n = \dfrac{111}{184}$

3. $a = 2\dfrac{1}{4}$ ft

4. You should do 38 problems.

5. Handle: $\dfrac{7}{10}$ feet; pole: $4\dfrac{9}{10}$ feet

6. Smaller length: $3\dfrac{3}{8}$ cm; larger length: $7\dfrac{17}{40}$ cm

Objective 1

1. $t = \dfrac{1}{14}$

3. $x = -\dfrac{3}{5}$

5. $n = -5\dfrac{2}{5}$

7. $y = 1\dfrac{13}{14}$

Objective 2

9. $\dfrac{1}{3}n = 25\dfrac{2}{3};\ n = 77$

11. $h - 2\dfrac{7}{10} = -4\dfrac{3}{5};\ h = -1\dfrac{9}{10}$

13. $\dfrac{1}{6}(a + 3) = 1\dfrac{1}{9} + a;\ a = -\dfrac{11}{15}$

Objective 3

15. $3\dfrac{1}{2}$ ft.

Objective 4

17. Each cooler holds 24 fl. oz.

Chapter 6

Section 6.1

Key Terms

1. fractions   2. ten-thousandths   3. left; right   4. 4

Practice

1. $\dfrac{7}{20}$

2. one hundred five thousandths

3. one hundred twenty-five and thirty-seven hundredths

4.

5. $>$

6. 21.063

Objective 1

1. $\dfrac{7}{8}$

3. $12\dfrac{1}{5}$

Objective 2

5. seven hundred eighty-nine hundred-thousandths
7. negative four and eighty-three hundredths

Objective 3

9.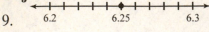

11.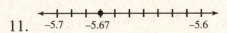

Objective 4

13. >

15. <

Objective 5

17. 21.0628

19. 21

Section 6.2

Practice

1. 130.7493
3. -48.506
5. $1.34a^4 + 0.056a^3 - 0.58a^2 + 1.34$
7. The missing side length is 21.2 mm.

2. 24.09
4. $12.4d + 5.4$
6. $x = -15.69$

Objective 1

1. 77.818

3. 167.6661

Objective 2

5. 78.21

7. 3.471

Objective 3

9. -134.3

11. 13.46

Objective 4

13. $4.3n + 3.2$

15. $0.02x^5 + x^3 - 0.02x^2 + 0.5x + 0.01$

Objective 5

17. $a = -109.162$

Objective 6

19. $757.23

Section 6.3

Practice
1. 2.0657
2. -16
3. -0.001
4. 267,000,000,000; two hundred sixty-seven billion
5. 7.89545×10^9
6. $8.68m^9n$
7. $0.6n^2 - 17.98n - 0.6$
8. $42.55

Objective 1
1. 4289.4785
3. 3.96534

Objective 2
5. -0.429
7. -37.03

Objective 3
9. 3.375
11. 0.0016

Objective 4
13. 58,970,000; fifty-eight million, nine hundred seventy thousand

Objective 5
15. 3.015715×10^{11}

Objective 6
17. $20.06m^3n^3$
19. $42.444r^3s$

Objective 7
21. $5.4x^6y + 4.23x^3y^2 - 0.18x^2y$
23. $x^2 - 33.32x + 26.65$

Objective 8
25. 14.44 m^2

Section 6.4

Key Terms
1. integers
2. rational numbers; irrational numbers
3. whole numbers
4. decimal numbers

Practice
1. 0.015
2. 20
3. $0.8\overline{3}$
4. 18.71
5. $2.8x^4$
6. $b = -8.24$
7. You need 3.454 ft. of tape.

Objective 1

1. 0.408   3. 3240   5. 2500

Objective 2

7. 14.5   9. −5.4

Objective 3   **Objective 4**

11. 0.25   13. $5x$

Objective 5

15. $x = 7$ 17. $z = -6.85$

Objective 6

19. $7.86

Section 6.5

Practice

1. 1.99   2. $0.5\overline{3}$
3. Jonathan's employer will reimburse him $580.33.
4. 2.73   5. 4.23×10^{21}
6. 10.5 cm^2 7. 10 in.^{2}
8. 34.1946 ft.^{2} 9. 23.55 m^3
10. 5.635 m^3 11. 60.3508 yd.^{3}
12. 2224.96632 mm^3 (however rounded)   13. 17.625 cm^2

Objective 1

1. −339.95   3. −11.18

Objective 2

5. 3   7. 26.2

Objective 3   **Objective 4**

9. The mean is 85.8.   11. 3.25

Objective 5

13. 480

Objective 6

15. 30.26 cm^2 17. 46.55 in.^{2}

Objective 7

19. 321.536 mi.^{3} 21. 2906.855 m^3

Objective 8

23. 83.4923 m^2

Section 6.6

Key Terms

1. hypotenuse
2. sum

Practice

1. $y = 1$
2. $x = -0.\overline{3}$
3. $h = 1.25$
4. It is about 93.338 ft. from home to second base.
5. Alex bought 10 CDs and 4 DVDs.

Objective 1

1. $y = -1$
3. $x = 1.5375$

Objective 2

5. $m = 0.05625$

Objective 3

7. $n = 42$

Objective 4

9. Holly can use her phone 152 feet into the backyard.

Objective 5

11. He sold 43 small cones and 15 large cones.

Chapter 7

Section 7.1

Key Terms

1. rate
2. unit
3. favorable; possible
4. quotient

Practice

1a. $\dfrac{7}{3}$ b. $\dfrac{18}{143}$
2. $\dfrac{1}{4}$
3. ≈ 17.37; The stock is selling at $17.37 for every $1 of annual earning
4. $1.77/lb.
5. 3 bars of brand A

Objective 1

1. $\dfrac{25}{42}$

Objective 2

3. $\dfrac{2}{13}$

5. $\dfrac{1}{2}$

Objective 3

7. ≈ 0.42; They owe about $0.42 for every $1 of income.

Objective 4

9. $\approx \$0.25\,/\,\text{ounce}$

Objective 5

11. half a pound of brand A deli meat

Section 7.2

Key Terms

1. congruent
3. cross products

2. proportional
4. similar

Practice

1. yes
3. 52 apples (will have 50 cents remaining)

2. $q = 35$
4. 150 m

Objective 1

1. no

3. yes

Objective 2

5. $n = 18.\overline{6}$

7. $x = 25$

Objective 3

9. 1.2 gal.

11. $24.26

Objective 4

13. $b = 7.2$ ft.; $c = 9.6$ ft.

Section 7.3

Practice

1. 48 in.
3. 24 pt.
5. 5 hr.

2. 657 ft.^{2}
4. 64 oz.
6. 1,381,320 mi./hr.

Objective 1

1. 5 ft.

3. 39,600 ft.

Objective 2

5. 0.11 yd.^{2}

Objective 3

7. $9\dfrac{3}{8}$ pt.

Objective 5

11. 0.164 yr.

Section 7.4

Practice

1. 730 mm
3. 800 g

Objective 1

1. 2300 m
5. 40,000 dm

Objective 2

7. 0.0001 kl
11. 15 ml

Objective 3

13. 890 mg
17. 60 dg

Objective 4

19. 1.7 cm^2

Section 7.5

Practice

1. $\approx$ 42,179 m
3. $\approx$ 121 lb.

Objective 1

1. 215.9 mm

Objective 2

5. $\approx$ 177.4 ml

Objective 3

9. $\approx$ 12.0 oz.

Objective 4

13. $-28.3°$ C

Objective 4

9. $277\dfrac{1}{3}$ oz.

Objective 6

13. $95.\overline{3}$ feet per second

2. 780 cl
4. 0.000038 km^2

3. 0.73 hm

9. 6000 L

15. 25 cg

21. 130 dam^2

2. $\approx$ 33.3 L
4. 38.3° C

3. $\approx$ 780.19 mi.

7. $\approx$ 70.4 L

11. $\approx$ 28.4 g

15. 176.7°C

Section 7.6

Key Terms
1. interest

2. back-end

Practice
1a. 0.19   b. 0.30   c. yes

2. $1344

3. $1767.10

4. 5 drops/min.

Objective 1
1a. 0.25   b. 0.44   c. no

Objective 2
3. $1334

Objective 3
5. $1533.60

Objective 4
7. 2 tablets

Chapter 8

Section 8.1

Key Terms

1. left

2. right

Practice
1. $\dfrac{2}{5}$

2. 0.95

3. $77.\overline{7}\%$

4. 0.4%

5.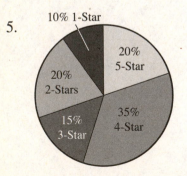

Objective 1

1. $\dfrac{3}{4}$

3. 0.048

Objective 2
5. 56%

7. 83%

Objective 3

9.

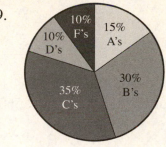

Section 8.2

Key Terms

1. is   2. what   3. %   4. of

Practice
1. whole
2. $0.9 \cdot 72 = n$; 64.8
3. $36\% \cdot n = 23.4$; 65
4. $n \cdot 40 = 16$; 40%

Objective 1
1. part

3. percent

Objective 2
5. $0.45 \cdot 120 = x$; 54

Objective 3
7. $0.4 \cdot w = 62$; 155

Objective 4
9. $n \cdot 81.5 = 32.6$; 40%

Section 8.3

Practice
1. $0.10 \cdot 220 = n$; 22
2. $28.50 = 0.68 \cdot n$; $42.00
3. $32 = n \cdot 33$; $96.\overline{96}\%$

Objective 1
1. $0.028 \cdot 79,000,000 = n$; $2,212,000

Objective 2
3. $0.30 \cdot n = 424$; $1413.33

Objective 3
5. $n \cdot 42 = 14$; $33\frac{1}{3}\%$

Section 8.4

Key Terms
1. new amount, original amount

Practice
1. Tax: $71.13; Total price: $1209.13
2. Amount of discount: $17.40, Final Price: $40.60
3. $16,000   4. 29.4%   5. Rate of discount: 20%; sale price: $560

Objective 1
1. $27.00

Objective 2
3. Amount of discount: $22; final price: $198

Objective 3
5. $26.75

Objective 4
7. 3%

Section 8.5

Key Terms
1. simple interest   2. compound interest   3. amortization

Practice
1. $18   2. Interest: $3000; Total: $13,000
3. $3896.76   4. $3810.60

Objective 1

5.

	Principal	Rate of interest	Time	Simple interest
1.	$1200	3%	1 year	$36
3.	$25,000	$5\frac{3}{8}\%$	$\frac{1}{4}$ year	$335.94

Total Amount: $2435.51

Objective 2
7. $3514.98   9. $5016.46

Objective 3
11. $1042.07   13. $3431.19

Chapter 9

Section 9.1

Key Terms
1. perpendicular   2. acute   3. ray   4. parallel   5. straight
6. complementary   7. congruent   8. interior

Practice
1. plane N 2. straight
3. congruent; They are vertical angles.   4. 50°, 130°, 50°, 130°
5. right triangle; The measure of one angle is 90°.

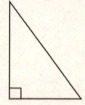

6. 38°, 41°, 101°   7. Not congruent
8. 50°, 38°, 92°

Objective 1
1. point O 3. line segment $\overline{XY}$

Objective 2
5. Right

Objective 3
7. congruent; They are alternate interior angles.
9. 116°   11. 55°

13.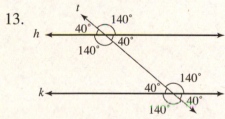

Objective 4
15. acute triangle; The measure of each angle is less than 90°.

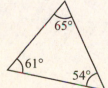

17. $72°, 52°, 56°$

Objective 5

19. Not congruent

21. $18\dfrac{2}{3}°$, $80\dfrac{2}{3}°$, $80\dfrac{2}{3}°$

Section 9.2

Key Terms

1. quadrants   2. left   3. coordinate   4. axis

Practice

1. $A(-2,4)$, $B(0,-3)$, $C(4,1)$, $D(-5,-3)$

2.   3. I   4. $(5,1)$

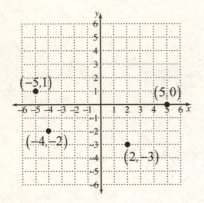

Objective 1

1. $A(1,3)$, $B(2,-4)$, $C(-4,-4)$, $D(3,-2)$

Objective 2

3.

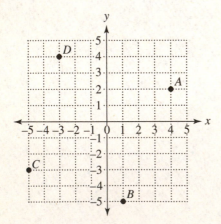

Objective 3

5. III   7. IV

Objective 4

9. $\left(2,-8\dfrac{1}{2}\right)$

11. $\left(1\dfrac{1}{2},0\right)$

13. $(-3.95, 2.05)$

Section 9.3

Practice

1. yes

2. $(0,-6),\ (1,-3),\ (2,0)$

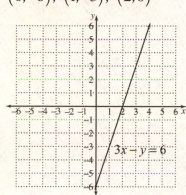

3. $(-4,6),\ (0,3),\ (4,0)$

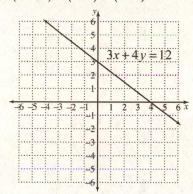

4. $(-3,4),\ (-3,6),\ (-3,0)$

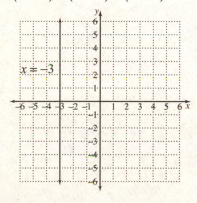

5. $(3,0);\ (0,2)$

Objective 1

1. no

3. no

Objective 2, Objective 3

5. $(-5,0)$, $(0,-5)$, $(-2,-3)$

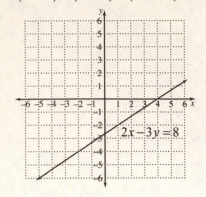

7. $(-6,6)$, $(0,0)$, $(6,-6)$

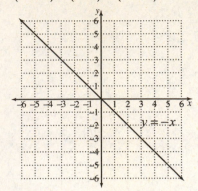

9. $(0,-5)$, $(1,-3)$, $(3,1)$

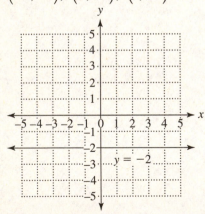

Objective 4

11. $(-3,-2)$, $(0,-2)$, $(1,-2)$

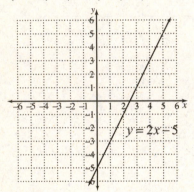

Objective 5

13. $(12,0)$; $(0,-8)$

15. $\left(\dfrac{25}{2},0\right)$; $\left(0,-\dfrac{5}{4}\right)$

17. $(0,0)$; $(0,0)$

19. $(-5,0)$; no y-intercept

Section 9.4

Key Terms
1. balance   2. mean

Practice
1a. 74 ft./sec.   b. 1.55 ft./sec.   c. ≈ 2.3 sec.

d.   2. $(-1.5, 2)$ 3. 40 square units

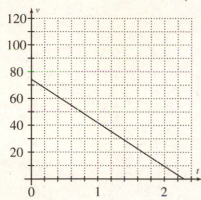

Objective 1
1a. $200   b. $373.60   c. $503.80

d.

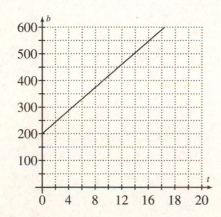

Objective 2
3. $(-6, -3)$

Objective 3
5. 33 square units

